A. H. Rubenstein, H. Schwärtzel (Eds.)

Intelligent Workstations for Professionals

Proceedings of a Joint Symposium

Siemens AG
Northwestern University

March 1992

With 90 Figures

Springer-Verlag Berlin Heidelberg GmbH

Professor Albert H. Rubenstein, PhD

Center for Information and Telecommunication Technology
Northwestern University

Professor Dr. Heinz Schwärtzel

Corporate Research and Development, Systems Technologies,
Siemens AG

ISBN 978-3-662-07956-0   ISBN 978-3-662-07954-6 (eBook)
DOI 10.1007/978-3-662-07954-6

Springer-Verlag Berlin Heidelberg 1993
Originally published by Springer-Verlag Berlin Heidelberg in 1993

Preface

Physicians, lawyers, engineers, architects, financial analysts, and other professionals articulate an increasing need for support by intelligent workstations for decision making, analysis, communication, and other activities. "Intelligent Workstations for Professionals" is the collection of papers presented by international scientists at a symposium and workshop in March 1992. Requirements from potential users, studies of their behavior as well as approaches and aspects of technical realizations of "intelligent" functions are introduced.

Eight contributions from members of the Center for Information and Telecommunication Technology (CITT) of Northwestern University, Wisconsin-Whitewater University, and the Children's Memorial Hospital deal with the latest findings of the UNIS (Users' Needs for Intelligent Systems) project, which is designed to identify needs and wishes from professionals for intelligent support systems and the potential barriers to adoption and use of such systems.

The remaining papers concentrate on new approaches and techniques that enhance the "intelligence" of future workstations. They tackle issues like architectural trends in workstation design, the combination of workstations with HDTV and speech processing, automatic reading and understanding of documents, the automated development of software, or the processing of inexact knowledge. These papers were contributed by members of the DFKI GmbH (German Research Institute for Artificial Intelligence), GMD mbH (German Society for Mathematics and Data Processing), Siemens Gammasonics Inc., Siemens Nixdorf Informationssysteme AG and Siemens AG.

Based on the broad spectrum of aspects treated in this book we feel that it may be suitable for research and development, sales and marketing, and for ambitious users of workstations who want to have "early warning" of features that will be common in future systems.

Evanston - Munich, August 1992   The Editors

Addresses

Deutsches Forschungsinstitut für Künstliche Intelligenz GmbH (DFKI)
Erwin-Schrödinger-Straße, 6750 Kaiserslautern, FRG

Gesellschaft für Mathematik und Datenverarbeitung mbH (GMD)
Schloß Birlinghoven, 5205 St. Augustin 1, FRG

Northwestern University
2145 Sheridan Road, Evanston, Illinois 60208, USA

Siemens AG
Otto-Hahn-Ring 6, 8000 München 83, FRG

Siemens Gammasonics Inc.
2501 North Barrington Road, Hofman Estates, Illinois 60195, USA

Siemens Nixdorf Informationssysteme AG
Otto-Hahn-Ring 6, 8000 München 83, FRG

The Children's Memorial Hospital
2300 Children's Placa, Chicago, Illinois 60614, USA

University of Wisconsin-Whitewater
Whitewater, Wisconsin 53190, USA

Table of Contents

AN INTRODUCTION TO THE UNIS PROJECT

(USERS' NEEDS FOR INTELLIGENT SYSTEMS)

Albert H. Rubenstein
Center for Information & Telecommunication Technology
Northwestern University

The major objectives and thrust of the UNIS project (Users' Needs for Intelligent Systems) are presented in the following paper by Geisler and Rubenstein. In this brief introduction, some "Spin Off" projects that developed out of the main thrust are described. Not all of them are represented by papers given at the Workshop/Seminar in Munich in March, 1992 and included in these proceedings. Approximately half of them are. Papers on the others have been written or are in preparation at Northwestern University.

As we began to probe into the UNIS area, a number of specific sub-issues emerged that required a closer look. In addition, the nature of the research team, about 15 members, provided opportunities to examine related questions that appeared to have potential for contributing to our knowledge of the general area as well as the specific aspects on which UNIS is focussed.

As a consequence, we have encouraged members of the team to explore some of these issues in parallel with the main stream project efforts described in the following papers. These additional explorations or "spin-off" projects are described briefly below.

2

A. DEGREE OF CUSTOMIZATION/STANDARDIZATION OF INDIVIDUAL INTELLIGENT ASSISTANTS THAT IS "REALLY" NEEDED AND IS COMMERCIALLY FEASIBLE

It became clear early in the study that we had to know something about the degree of customization/standardization of potential "Intelligent Machines" or "Intelligent Assistants" that might be cost-effective for both the users and the vendors of such systems. The paper by Nagaraja Srivatsan explores the views of vendors, information specialists and users on this issue.

B. AN INTELLIGENT INFORMATION RETRIEVAL SYSTEM FOR SUPPORTING CLINICAL DECISION MAKING IN MEDICINE

The UNIS project enabled us to "piggy-back" on a project Dr. M. Paul had been carrying on for many years in connection with his own clinical and research work in pediatric cardiology and in the general area of his interest in medical informatics. In turn, involvement with UNIS has led to further development in his ICARIS research and the possibilities of demonstrating the system in the near future.

C. THE ROLE OF INTUITION IN MANAGING TECHNO-LOGY/R&D

Dr. Milton Glaser -- a retired Vice President of Research and Development in a chemical company -- has long held an interest in the role of intuition in decision-making in R&D. Involvement in the project team has enabled him to pursue that interest in a systematic study of the intuitive components of an R&D manager's job.

D. AN INTELLIGENT SOFTWARE DEVELOPMENT WORK STATION

By coincidence, Dean Koester, an employee of Siemens/Gammasonics in the Chicago area, has been pursuing graduate work at Northwestern in the field of Engineering Management.

As part of that work and his own continuing interest in the software development process, he has focussed his efforts on the needs for and characteristics of software development workstations.

E. FACTORS AFFECTING RESISTANCE TO INTEREST IN AND ACCEPTANCE OF INTELLIGENT COMPUTER ASSISTANCE IN HEALTH CARE

This work by Jim Felli, a Ph.D. candidate in the IE/MS department at Northwestern, is represented by a paper in these proceedings and combines both main stream work on UNIS and his additional personal interests in the use of computers in the health care field.

F. INTELLIGENT SYSTEMS FOR THE AEC (ARCHITECTURE, ENGINEERING, CONSTRUCTION) INDUSTRY

Since he was not able to attend the Munich seminar, Tom Donzelli's work in this field is not included in the proceedings. As a spin-off of the main project and reflecting his own experience in the AEC industry (architecture, engineering, and construction) he has looked into the opportunities for future introduction of intelligent systems and other advanced expert systems in this highly fragmented and, historically, "low technology" field.

G. THE ROLE OF VOICE MESSAGING AND OTHER "GROUPWARE" TECHNIQUES IN INTELLIGENT NETWORKS SUPPORTING PROJECT MANAGEMENT

The paper by Ms. Rhea Walker, a Ph.D. candidate in Communication at Northwestern who is a member of the project team, reflects what has become a major spin-off from the project. This is a series of studies on electronic support for communication and decision-making -- often called GROUPWARE. Her many years of experience in the telecommunication industry have prepared her to lead a team in our new research thrust in this area.

H. AN INTELLIGENT ASSISTANT FOR SUPPORT OF THE CHIEF TECHNOLOGY OFFICER (CTO) IN A HOSPITAL

Although not included in these proceedings, the work on this topic has also developed into a major research thrust at Northwestern through the personal interest of Dr. Ori Heller, a physician who is pursuing the idea that hospitals and other health care organizations require major coordination and leadership in their acquisition and use of technology throughout the organization.

I. THE ROLE OF THE INFORMATION SYSTEM DEVELOPER VIS-A-VIS POTENTIAL USERS IN DEVELOPMENT OF INTELLIGENT SYSTEMS -- HOW MUCH USER INVOLVEMENT

Dr. Gerald Hoffman's paper in this proceedings is on a main theme in the UNIS project -- the components or "engines" required to drive future intelligent systems for the professional. The topic of "User Involvement" is intellectually related and was a spin off of his involvement in the main project, but is being pursued as a separate research project in a number of organizations in the service sectors and with collaborators in Canada.

J. ADDING INTELLIGENCE TO THE INFORMATION AND ANALYSIS SYSTEMS IN A MEDICAL LABORATORY

This project, being pursued by Peter Lai a Ph.D. alumnus of Northwestern, is currently on hold due to changes in direction by our potential industrial collaborator. It is a logical extension of the UNIS work, since it relates to the interface between health care professionals and the equipment and systems that generate data and other information needed for clinical decision-making and other purposes.

This is not a complete list of all the spin-offs from the UNIS project, but it is representative of how that project has opened up entire new areas of research within the Center for Information and Telecommunication Technology at Northwestern University. Other areas we are pursuing relate to direct spin-offs in individual professions -- e.g. the

Law Office of the Future, the Hospital of the Future with Respect to Information Technology, The Load on Corporate Networks created by the introduction of an increasing array of new communication and decision-support systems, the need for improved imaging and multi-media capabilities, and others.

USERS' NEEDS FOR INTELLIGENT SUPPORT SYSTEMS (UNIS): A STUDY OF POTENTIAL ADOPTION BY PROFESSIONALS

Albert H. Rubenstein
Center for Information and Telecommunication Technology
Northwestern University

Eliezer Geisler
University of Wisconsin-Whitewater

1. INTRODUCTION

A three year project sponsored by Siemens AG began in 1990, aimed at the study of needs of professionals for intelligent support systems and the potential adoption of such systems. In this article the authors describe the objectives of the study, the methodology employed, and initial findings from the first phases of the study. They also suggest some strategic as well as marketing implications for potential vendors of future intelligent support systems.

2. PREVIOUS STUDIES OF INTELLIGENT SUPPORT SYSTEMS

Intelligent systems are generally defined as artificial systems displaying behavior which could be considered intelligent. Different definitions exist in the various literatures. When discussing technical aspects of intelligent systems, the terms "expert systems," "knowledge-based systems" and "artificial intelligence" are used frequently. The business literature often uses descriptors such as "computer systems with capabilities similar to those of human decision makers." Such terminology also includes "decision-support systems" and "executive-support systems."

Early stage intelligent systems are being used for many purposes in business, science and engineering and other areas. The issues of interest to us in this project are the interactions between potential future intelligent systems and professionals such as physicians, lawyers, financial analysts, engineers, architects, and other professionals.

In the medical profession, for example, there are intelligent systems for diagnostics, such as: Al/Rheum for rheumatology; Mycin for infectious diseases, and Al/Gen for hearing and sight loss diseases. Research trends in medicine include the development of intelligent networks between hospitals and individual physicians, which accommodate information technology, and expert systems for diagnostics and consulting on treatment and related matters.

In the financial industry, insurance companies utilize such systems as USER (Underwriting, Strategy and Reporting) and PACE (Procedures for Automated Claims Examining). Banks utilize a variety of systems for check processing, accounts resolution, loan management and customer service. Research trends in banking include comprehensive imaging techniques, automated pattern recognition, talking credit cards and improved networking.

Key issues in the interaction between humans and these intelligent systems have not been well researched nor solved by firms which develop and utilize these systems. Some studies explore user behavior patterns such as how people input and extract data –– mainly the person's physical interaction with computers. Other studies employ psychological theory to explore decision-making processes and human-machine interaction. Current research trends are exploring contingent decision behavior and the choice of heuristics.

In science and engineering, research trends emphasize interaction with new kinds of computers which are rapidly emerging (general multiprogrammed supercomputers, monoprogrammed computers, and application-specific computers). Other areas include distributed supercomputing and targeting higher performance machines for various tasks.

Although a substantial body of literature is emerging on uses of expert systems and other intelligent support systems, there has been a need for systematic and broad studies of the **needs** for and potential **utilization** of "next generation" intelligent systems. Such studies involve the exploration of user behavior patterns and their interaction with both existing and potential new high performance intelligent systems.

3. OBJECTIVES OF THE PROJECT

This project is a theory-based empirical study of the needs of professionals such as physicians, lawyers, engineers, architects, and financial analysts for intelligent computer and telecommunications support in their work-related activities. It deals with the thought processes and behavior of these professionals in areas such as: analysis, information searching and using, planning, strategy and scenario development, decision making, and evaluation.

The *first objective* of this pre-commercialization study is to improve our understanding of the work flow of the professional and his/her organization. The theoretical base for this first objective is being developed through a series of propositions, extracted from interviews with professionals, about: the way in which they do their work, the problems they encounter, and their needs for new and improved intelligent support for that work.

The *second objective* is to identify opportunities for new and enhanced electronic support in the form of software, systems, and "machines" or "intelligent assistants" that can help make these professionals more productive and effective in accomplishing their organizational and professional goals.

The *third objective* is to juxtapose such ideas for new machines with advancing technology and specific intelligent support systems which are currently being developed or conceived, to see what kinds of matches might be made between such users' needs and the kinds of products and services that might be offered in the next 5-10 years.

The following research questions guide the study: a) what are professionals' potential needs in the future for intelligent support systems?; b) what factors are conducive to the potential adoption of such systems?; c) what factors delay or prevent professionals from adopting and utilizing intelligent support systems?; d) what factors (in b and c) are specific to certain professions, and why?; and e) is there a threshold at which professionals turn to the use of intelligent systems?

4. METHODOLOGY

The project, now in its third year, has three phases. In the first phase our objectives were to develop, through in-depth interviews with professionals, the following: a) a list of likely future needs for intelligent assistance in their professional tasks; b) an initial list of "machines" or "intelligent assistants," and c) an initial list of propositions about the work of the professional in relation to the potential adoption and use of intelligent assistance.

The second phase, which began in 1992, consists of "real time" data collection of professionals' needs and potential adoption of intelligent support. This is done by continuous self-monitoring and reporting by professionals at work, followed by interviews and "debriefing" by our project team. The objective is to identify a number of "events" in the work of the professional which might evoke the need for intelligent assistance.

The third phase consists of a potential field experiment in which selected prototypes or simulations of intelligent support systems are provided to professionals. We then plan to study the potential adoption process, including factors influencing resistance to trial, adoption, and use. To date we have studied samples of: physicians, lawyers, architects, engineers, financial analysts, and R&D managers.

5. ACCOMPLISHMENTS IN THE FIRST PHASE

We reviewed the relevant literature for existing instruments and methods in the area of adoption and implementation of expert systems and potential intelligent support systems. We also reviewed the relevant literature on professionals and human-machine interaction. We conducted a series of seminars and group meetings of our "UNIS" project team, resulting in over 35 "idea memos" circulated among the more than a dozen members of the project team. Further, we developed two conceptual schemes which provide background and guidance to our study. The first is shown in Figure 1. The figure details some job-related activities of professionals and the possible benefits these may derive from intelligent support. A second conceptual scheme is shown in Figure 2. This figure identifies factors which may influence the potential adoption and usage of intelligent support systems. Such factors range from the impact of vendors and the environment of the organization to individual and professional variables.

In phase I, as of July 1992, we interviewed 38 professionals in 10 organizations. We then extracted from the interviews 28 ideas for future or potential "intelligent assistants" or "machines" and over 100 propositions on the behavior and perceptions of professionals and their organizations. The following is a partial list of the "machines" and their potential benefits:

1) *"Disaster backup" machine*: How to recover if the computer breaks down due to power, telcom or other failure outside the firm. *Potential benefits*: Avoidance of disaster by saving data, thus saving both direct and indirect costs in time and money and "lost opportunities,"

2) *"Predicting future evolution of the law in a particular area" machine*: e.g., how to recommend to clients purchase of real estate without assuming specific pollution cleanup liabilities. *Potential benefits*: Increased competitiveness of the professional's organization.

3) *"Identification of the characteristics of a deal" machine*: This is an intelligent system which will associate the characteristics of a deal (e.g., a real estate or joint venture deal) with a specific category which will then lead to a given structure which, in turn, commands the preparation of such necessary elements/items as leasebook, option, sale material, technology exchange agreements, etc. *Potential benefits*: Speeding up process by an estimated factor of 3-5.

4) *"Assuring that all important documents/elements/items are incorporated in a file/deal" machine*: This is an intelligent system which will not only keep track of relevant documents but also assure that all crucial documents are included and that they include the latest knowledge (case law, etc.). *Potential benefits*: Speeding up the process, cutting the cost of human resources assisting in these tasks, and avoiding serious omissions.

5) *"Analysis of the predicted behavior and predicted reaction of opposing attorneys" machine*: This intelligent system would analyze and predict the behavior and the reactions of given attorneys, in specific situations, based on analyses of their court

experience (actions, pronouncements, questioning, etc.). *Potential benefits*: Reducing uncertainty, improving competitiveness, increasing odds of winning or successfully negotiating cases, improving service to client.

6) *"Alerting the professional to inconsistencies in documents" machine*: This intelligent system analyzes the document/case and alerts the professional to inconsistencies, when such inconsistencies are detected.   Although this machine was proposed by and for the legal profession, it has, perhaps, a universal appeal to any professional preparing a document (for example, students, academics, bank personnel, government workers, insurance personnel, etc.). *Potential benefits*: Improved speed, improved accuracy, improved effectiveness and competitiveness of the professional and his/her organization, avoidance of errors.

7) *"Comparison of changes in standards demanded by professional associations to changes demanded by client and/or the nature of the job" machine*: This is an intelligent system which compares the changes in standards demanded by professional associations with those introduced by the professional and/or suggested by clients. Examples are types of materials in construction projects. Although this machine was recommended for and by architects, it may be of general use to engineers, construction personnel and others who work with two similar (perhaps identical) databases, yet separated by time. Therefore, whenever standards are used, this machine would show the changes that are introduced by the professional, as compared with standards imposed by the professional association, government agencies, and other regulators and external entities. *Potential benefits*: Savings of time and other resources, improved competitiveness, avoidance of disasters or litigation by the inadvertent use of outdated standards.

8) *"Integration of various and sundry systems into one integrated intelligent system" machine*: This is an intelligent system which would incorporate into one integrated system the currently available systems of, for example: CAD, Financial data, Planning data, Administrative data, Personnel data, Standards, and Regulations –– all in one system with one screen on which one can call up different "windows" but with a *unified*, *simple*, and *user* friendly interface. As in previous systems, although recommended by architects for architects, this system may have a broader appeal to

12

many other professions in which technical, administrative, and regulatory data currently exist in different data bases and systems. Clearly there may be numerous human as well as organizational and psychological barriers to the implementation of such a system. Nevertheless, the appeal of an *integrative intelligent system* seems to be extremely strong. *Potential benefits*: Improved ability to perform on time and on budget, savings of time and resources, improved competitiveness.

9) *"Convergence and clarification" machine*: This intelligent system combines the views of different constituencies, such as clients, experts, and informants or parties to a conflict, into a converging viewpoint, by perhaps associating and combining those ideas and viewpoints on which the parties seem to be in some agreement. Although this machine was proposed by an engineer, who is a technical manager, it may have a general appeal to other professionals for conflict resolution, reduced time of negotiations, and reconciliation of divergent positions. *Potential benefits*: Improved effectiveness of the professional, time savings in meetings and negotiations.

10) *"Hopes and dreams extraction" machine*: This intelligent system identifies non-obvious or latent needs and potential needs, demands and trends from potential users or customers or clients. Although suggested by a technical manager, the machine may have a general appeal to professionals in marketing and advertising organizations, in public relations and public opinion organizations, and in product design and development. *Potential benefits*: Improved productivity, increased competitiveness, improved market research, and product/service planning.

In addition to the above, more than a dozen additional machines for physicians and other professionals are described in other reports on the UNIS study (see section 9). Important aspects of the potential intelligent machines are the components of such machines which will be needed to make them useful. To date, we have identified seven such components. They are listed in Figure 3. The technical feasibility of these components will depend on the development of extremely fast processors, very high capacity direct access storage devices, and data channels with very high throughput. We estimate that capacity improvements of two orders of magnitude will be required for some of these components, and that economic feasibility will require significant improvements in cost/performance.

In addition to the machines and their components, phase I also generated over 100 propositions on professionals' thought processes and their attitudes toward intelligent support systems. These are potentially testable propositions extracted from our interviews with members of these five professional groups. Figure 4 shows an illustrative sample of the propositions.

6. ACCOMPLISHMENTS IN THE SECOND PHASE

In phase II we have begun conducting "real time" data collection of the professional's needs for and attitudes toward potential adoption of intelligent support. To this end we designed a 3×5 card, which is very portable and easily fits in the respondent's shirt or jacket pocket. Respondents are asked to carry the card on their person and to complete the card *at the time* a need for intelligent support arises during their daily routine. The magnified card is shown in Figure 5. A total of 62 professionals received the card. As of July 1992, 18 returned the completed card, a 29 percent rate of return. (They include: 6 lawyers, 5 R&D managers, 2 physicians and 5 architects). The card included a short list of about a dozen "machines" suggested by respondents in phase I of the study. We provided this list primarily as guidance to the respondents on the kind of intelligent support we are considering.

Upon receipt of the card, we conduct "post-card" interviews. To date, twelve respondents were interviewed. A pilot group of four respondents was interviewed, with the objective to develop the interview protocol for "post-phase II" data collection. We have contacted the remainder of the respondents who completed the card, for interviews during the summer of 1992. These interviews are currently underway. Finally, we have also contacted those professionals who received the cards yet have not returned them. At the same time, we also sent phase II cards to a new sample of 38 professionals (for a total of 100 contacted in phase II), who tentatively agreed to participate in our study.

7.   SOME INITIAL FINDINGS

7.1   Findings from Phase I

1)   It is feasible to interact with professionals so that they are motivated to speculate on future needs for intelligent systems and, indeed, propose such highly speculative and futuristic machines. We found that it is not only feasible to probe the professional's mind for such speculations, but that this process can be an interesting and stimulating experience for the professional, who has the opportunity to sit back and to think about the future, independent of the pressing weight of current duties and responsibilities. This atmosphere of intellectual curiosity should greatly facilitate the planned work for the remainder of the project.

2)   Many of the machines suggested by the professionals we interviewed are geared towards assisting the professional in the mundane, routine, not-so-professional tasks. There seems to be a considerable amount of resistance to delegate the "truly" professional tasks to a machine. Thus the respondents propose advanced machines which would assist the professional to manage, administer, analyze and communicate. However, the "core" of the professional's task is seldom relegated to a machine, however futuristic and full of features and capabilities.   This resistance is perhaps related to the inherent qualities of the professional –– due to his/her training, style of operation (e.g., close interaction with the client or patient; see Rubenstein *et al.*, "Explorations on the Information Seeking Style of Researchers," in: Nelson and Pollock (Eds.); **Communication Among Scientists and Engineers**, D. C. Heath & Co., Lexington, MA, 1970), and psychological make-up.

3)   There seems to be convergence across professions, in that similar types of machines are suggested by different professionals. Partly this may be explained by the reluctance of professionals to delegate their "core" professional tasks to machines and the increasing preoccupation of professionals across disciplines with the tasks of administration and management of their offices and staffs.   Since administration and management are essentially composed of the same or similar functions and tasks, this leads to similarity in some of the machines proposed.

4)   The professionals we interviewed seem reluctant to link existing "expert systems" with their proposed advanced machines. That is, they do not think of "intelligent systems"

simply as extensions of existing expert systems, decision support systems, or executive support systems. Rather, they seem to intuitively distinguish between current expert systems and potential future intelligent systems.

7.2 Findings from Phase II

1. Professionals tend to relate the machines to *actual* occurrences in their professional activities. The attorneys related the machines to a court incident and to the preparation of a legal document. The physicians more often related the machines to their "office" activities rather than "patient treatment" activities.

2. The machines selected by the professionals we interviewed in phase II thus far are intelligent machines which produce shorter term results. That is, they prefer to choose machines such as: "money management machine" and "office management machine" and "inconsistency alert" rather than machines such as "hopes and dreams" machine. A possible explanation may be the relation the respondents perceive between *actual* occurrences and the need for intelligent support machines in the present situations.

3. It is feasible to investigate the relation between actual professional activities and the need for intelligent support machines by professionals -- in "real time."

4. Although our preliminary sample thus far is very small, we did find that professionals seem to have a clear idea of their need for intelligent support in *actual* work conditions. When professionals are made to think about their need for intelligent support, *at the time of such need*, they are more apt to visualize a needed intelligent support machine.

5. Interviews to date have revealed that:

   (a) Professionals we interviewed seem to extrapolate from existing machines to those they need -- by adding some features they require at that moment.

   (b) Major concern of the professional in choosing a needed machine seems to be savings in time rather than cost or other criteria.

6. Respondents were able to identify events/situations/cases and circumstances which led them to realize the need for ISS machines. Therefore, we are able to obtain a list of such events. In the few phone conversations with respondents after they completed the cards, we find that many such events were defined by them as "crisis" events, non-

routine in nature. These crises generated in the respondent the thought of needing some assistance and the thought of "perhaps I should put this down in the card."

7. The cards also reveal several such crises or events on a *single* date. This may prove important in our forthcoming interviews. The professionals who completed the cards have sometimes several crises or situations that may require ISS support -- in a single day or even in a single professional occurrence. Speculation is that any future ISS designed for the professional's use -- to be viable -- needs to address more than one need or have more than one capability to resolve the problems or needs arising in a single professional occurrence or case. This issue needs further investigation.

8. Findings across professions are still sketchy, due to the small number of cards returned. However, we can already say that there are preliminary indications that -- across professional boundaries -- we can relate, correlate and link the events/problems/cases to specific ISS machines (at least in the perceptions of respondents). That is, our methodology allows us to isolate a given event/problem/case which was extracted from a *real-time* experience of the professional -- and to discuss such an event in relation to ISS/machines. This fact is useful not only when we return to the professional and interview him/her about the card and their attitudes/feelings/needs, but also in creating a useful unit-of-analysis for phase II: "a real-time event *and* the appropriate machine related to it."

8. SOME IMPLICATIONS FOR POTENTIAL VENDORS

Our findings from phases one and two indicate that, in the near term future, intelligent support systems are likely to be adopted by the professionals we studied primarily for assistance in tasks which are more routine and well-focussed, and which include assembly, codification, categorization, and structuring of data within a given framework -- rather than in the deep analysis and interpretation of such data. In addition, our findings indicate that the intelligent systems more likely to be adopted are perceived to perform functions of *assisting* the professional in the conduct of his/her professional tasks, rather than in *replacing* the professional in the conduct of these tasks.

Both large and small vendors will be looking at a potential market for intelligent support for professionals, in which there is ample room for new products and services.

The products more likely to be adopted in the near term would be those geared towards clear and substantial improvements in assistance to the professional in more routine tasks. Assistance in highly conceptual tasks requiring synthetic reasoning would become attractive to professionals only in the longer term of perhaps 10-15 years -- after the above-described "machines" had been successfully implemented and become "routine."

9. PAPERS AND PUBLICATIONS FROM THE UNIS PROJECT TO DATE

Donzelli, Tom, "The AEC Industry," UNIS Working Paper, January 10, 1992 (92/09).

*Felli, Jim, "Why Is There Such Limited Use of Intelligent Systems in Medicine," March, 1992.

Geisler, E., and Rubenstein, A. H., "Factors Influencing the Adoption and Use of Intelligent Systems in Professional Service Organizations," *Proceedings of the Portland International Conference on the Management of Technology*, October 27-31, 1991, Portland, Oregon (90/64).

Geisler, E. and Rubenstein, A. H., "Barriers to the Adoption of Intelligent Systems by U. S. Lawyers," forthcoming in: *International Journal of Computer Applications in Technology*, October 1991 (91/14).

*Geisler, E., and Rubenstein, A. H., "Users' Needs for Intelligent Systems: Some Findings from an Empirical Study of Professionals in Service Organizations," March, 1992.

Hoffman, Gerald, "Dreamware: Intelligent Systems to Support the Managers of the 1990s," CITT, January, 1991 (91/03).

Koester, Dean, "The Software Development Machine," January 1992, CITT (92/04).

Koester, Dean, "Software Development Machine," UNIS Working Paper, 26 February 1991 (91/06).

*Paul, M., and Goldberg, P., "ICARIS - Integrated Clinical and Research Information System," March, 1992.

Rubenstein, A. H., "Some Spin-Off Projects from the Study of Users' Needs for Intelligent Systems (UNIS), CITT, September 1991 (91/48).

Rubenstein, A. H., "Users' Needs for Intelligent Systems: First Year Progress Report," CITT, April 1991 (91/02).

---

*The articles with asterisks are included in: Siemens, A. G., **Intelligent Workstations for Professionals: Collection of Slides**, Symposium/Workshop, March 23-24, 1992, Munich, Germany (92/55).

Rubenstein, A. H., Geisler, E., *et al.*, "Preliminary Design for a Series of Intelligent Support Machines for Professional Knowledge Workers," *Proceedings of the 3rd International Conference on Management of Technology*, Miami, Florida, February 17-21, 1992 (91/15).

Rubenstein, A. H. and Geisler, E., "Users' Needs for Intelligent Systems: A Panel," prepared for the TIMS/ORSA Meeting, Anaheim, California, November 3-6, 1991 (91/10).

Sirimongkolkasem, J., "Groupware," UNIS Report, 3 December 1991 (92/15).

Srivatsan, N. and Hoffman, G., "System Components of Intelligent Machines," CITT, May 1991 (91/23).

*Srivatsan, N., "Customization/Standardization Issues in Expert Systems Development/ Marketing," UNIS Paper, June, 1991 (91/28).

Srivatsan, Nagaraja, "Intelligent Systems: New Product Development for the Future," March 1992.

Walker, Rhea, "A Review of the Literature on Intelligent Systems: Their Impact on Professional and Service Business Sectors," CITT, December 1989 (90/28)

*Walker, Rhea, "Groupware," March, 1992.

ACKNOWLEDGMENTS

We appreciate the contributions to this paper of the following members of our UNIS project team:

Gerald Hoffman (The Gerald Hoffman Company); Milton Paul, M.D. Northwestern Medical School); Milton Glaser (Dexter-Midland, retired); Dean Koester (Siemens Gammasonics); Rhea Walker (AT&T and CITT); Nagaraja R. Srivatsan, Cathy Wilkes, James Felli, Tom Donzelli, Gary Summers (all with CITT); Jamaluddin Husain (Purdue University-Calumet); Ori Heller, M.D. (Israel Defense Forces); and Peter Lai (Northeastern Illinois University).

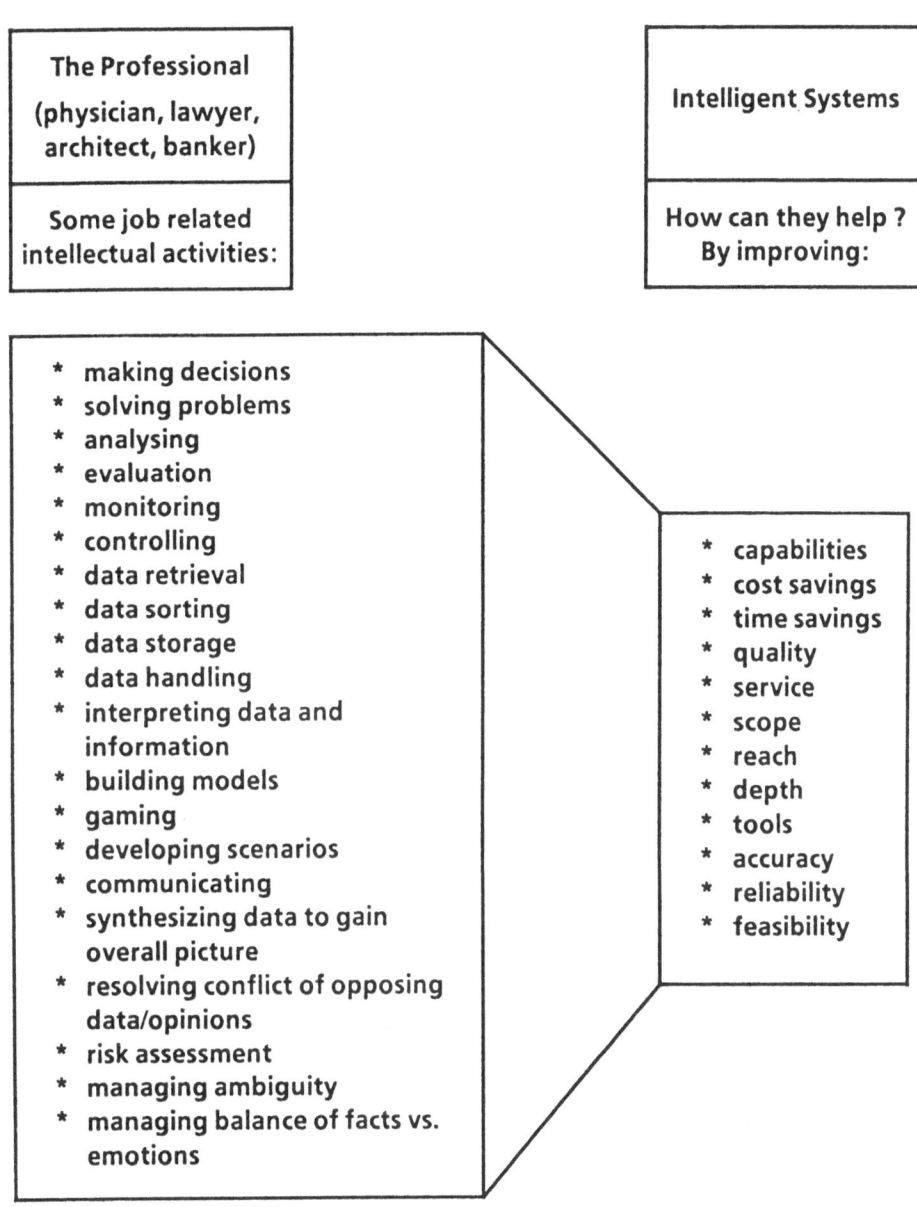

The Professional
(physician, lawyer,
architect, banker)

Some job related
intellectual activities:

Intelligent Systems

How can they help ?
By improving:

* making decisions
* solving problems
* analysing
* evaluation
* monitoring
* controlling
* data retrieval
* data sorting
* data storage
* data handling
* interpreting data and
  information
* building models
* gaming
* developing scenarios
* communicating
* synthesizing data to gain
  overall picture
* resolving conflict of opposing
  data/opinions
* risk assessment
* managing ambiguity
* managing balance of facts vs.
  emotions

* capabilities
* cost savings
* time savings
* quality
* service
* scope
* reach
* depth
* tools
* accuracy
* reliability
* feasibility

Figure 1: Very preliminary conceptual scheme of the professional's activities.

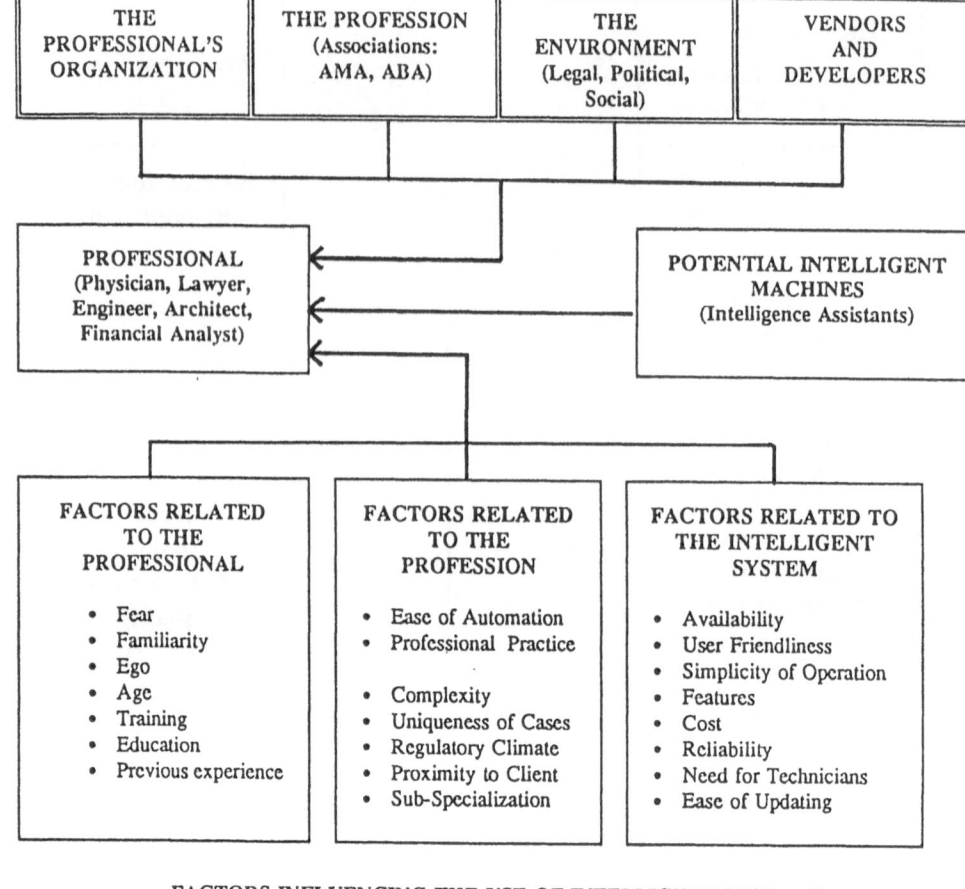

FACTORS INFLUENCING THE USE OF INTELLIGENT SYSTEMS

RESEARCH QUESTIONS:

1. *What factors are conducive to the use of intelligent systems by a professional?*
2. *What factors are specific to certain professions, and why?*
3. *At what point do professionals turn to the use of intelligent systems? Is there a threshold?*
4. *What factors delay or prevent the professional from utilizing intelligent systems?*

Figure 2: Very preliminary conceptual scheme: user's needs for intelligent systems by professionals.

FIGURE 3: SOME COMPONENTS REQUIRED FOR POTENTIAL INTELLIGENT SUPPORT SYSTEMS

1) **Semantic/Syntactic Analyzer (SSA)**

The SSA extracts meaning from text by analyzing both syntax and semantics. Its output is a set of assertions in some small number of standard formats suitable for further analysis and manipulation. For example:

The probability that fact A is true is P.

Event B happened between time t and time u.

Data element C, required by process D, is missing.

2) **Rule Maker (RM)**

The purpose of RM is to create the rule base of an expert system without asking an expert explicit rule-oriented questions. It uses the output of SSA (based on whatever documents and other data is available) to create rules which can be used by an inference engine.

3) **Consistency Checker (CC)**

CC analyzes the output of SSA for logical inconsistencies. It resolves them internally insofar as it can and reports what it has done. It refers unresolvable contradictions to human operators.

4) **Trend Analyzer (TA)**

The purpose of the Trend Analyzer is to predict the evolution of events which are described verbally rather than numerically. It is analogous to statistical techniques such as Box-Jenkins, multiple regression, etc. It uses the output of SSA to predict the likelihood of future events. It analyzes trends and pinpoints outliers; it describes the consequences of both likely and unlikely outcomes.

5) **Options Generator (OG)**

OG uses the results of SSA (as modified by CC) to create declarative statements which can be understood by humans and can be used to drive word processors and other programs.

6) **Synergistic Information Collector (SIC)**

SIC examines user requests for information and extends the requests to related subjects, by identifying topics and concepts retrieved by the original request but which were not an explicit part of it. The process can be continued recursively to find 'topics related to related topics' as far as required.

7) **Redundancy Suppressor (RS)**

RS examines sets of documents, identifies redundant information, and reports non-redundant information together with memoranda describing suppressed materials.

Figure 3: Some components required for potential intelligent support systems.

FIGURE 4: ILLUSTRATIVE SAMPLE OF PROPOSITIONS

P1   The main goal of a lawyer is to get for his clients the best economic and legal results, taking into account the body of law.

P2   Much of a lawyer's job is done by instinct.

P3   The thought processes of a lawyer usually have two stages. First is the stage of analysis, followed by the stage of synthesis.

P4   If intelligent systems substitute for a large portion of a lawyer's professional tasks, then such automation will be conducive to "lazy" practitioners of the law, in a way similar to that of calculators preempting the acquisition of skills in elementary math in grade school children.

P5   Since the legal profession deals with almost every combination of events, legal maneuvering and legal factors, there are difficulties in channeling tasks in a logical and rule-based manner.

P6   The vast majority of physicians do not have a personal computer in their examining room or personal office, nor do they plan on introducing one.

P7   In most medical specialties the physician's thought process is quite straightforward, involving diagnosis and, once this is established, assignment of treatment.

P8   Since medical diagnosis is a pyramid of knowledge, the computer's role in the pyramid is at its top, once all the thinking has been done as background.

P9   If medical intelligent systems are to become useful and effectively used by physicians, they should be able to identify subtle indicators (symptoms) which exist in the patient but which are unknown to the patient who therefore does not tell the MD, leading perhaps to misdiagnosis.

P10 Acceptance of medical intelligent systems is incremental and very slow due to the need to fit the outcome (values) generated by new technology -- primarily new tests and diagnostic techniques -- into the existing framework of thinking and logical processes of the practicing physician.

Figure 4:   Illustrative sample of propositions.

23

P11 Intelligent systems in medicine will become more sophisticated and replace some physician's thought processes mainly because: 1) the profession is changing and many new MDs are of lower quality, 2) these new MDs tend to transfer the burden of their profession to machines, 3) they prefer a 9-5 workday, and 4) they may not be intelligent enough to practice medicine the way it is practiced today.

P12 Engineers are a type of professional who have a lot of enthusiasm but are not able to put into words their ideas, findings, and thought processes -- thus making automation of their tasks very difficult.

P13 Engineers know the degree of rigidity or toughness of examiners and other such controlling people and thus tend to "tailor" their documentation and arguments in line with the degree of toughness.

P14 Most lawyers cannot explain nor ta the "artistic" component of their thought processes, nor can they easily transfer such a component to an intelligent system.

P15 Systems development professionals find the most beneficial intelligent systems those systems which utilize a wide array of sources of information while providing their clients with answers which are concise, consistent and much richer in information as a result of the use of the machines.

Figure 4 (contd):   Illustrative sample of propositions.

24

FIGURE 5:   MAGNIFIED VERSON OF PHASE II CARD

PERIODIC DATA CARD

Please complete this card for the period of the next two weeks of your usual professional activities. For each event or problem, or case, please enter -- AT THE TIME THE EVENT OCCURS -- a possible intelligent support system or systems ("machine/s") that you think might help you to deal with the event, solve the problem, etc. Please feel free to list any of the "machines" lsied on the back of this card, or suggest a different "machine/s." *For any questions please call us at (708) 491-7928. Ask for Gary Summers or Dr. Elie Geisler.*

DATE	EVENT/PROBLEM/CASE	"MACHINE"
_____	_____	_____
_____	_____	_____
_____	_____	_____

STARTING DATE: _____   END-DATE: _____

LIST OF POTENTIAL "MACHINES" PROPOSED BY PROFESSIONALS

This list of machines was suggested by professionals we previously interviewed in your area. It contains some potential intelligent support systems which, if existing in today's marketplace, could have provided them some professional assistance in their work.

MACHINE NAME	POTENTIAL USES
1. "Disaster Backup Machine"	Recovery from computer breakdowns or failure outside firm.
2. "Behavior Analysis Machine"	Analyzes and predicts behavior/reactions of clients, competitors and others.
3. "Inconsistency Alert Machine"	Analyzes documents and alerts writer of inconsistencies.
4. "Convergence and Clarification Machine"	Combines and converges views of different parties.
5. "Hopes and Dreams Machine"	Identifies non obvious or latent needs of clients.
6. "Audio Hardcopy Machine"	Converts voice directly to hard copy outputs.
7. "Money Management Machine"	Estimates expected revenues, cash inflows, real-time costs of serving clients.
8. "On-Line Info Expert Machine"	Performs two tasks: a) assembles information/data from client, stores and adds to data base; b) responds directly to questions on data base by voice of professional or designee. Provides outputs in mode and format desired.
9. "Learning the Ropes/ Mentor Machine"	Assists newcomers in profession to learn the "ropes" and enculturates them to the professional organization.
10. "Office Management Machine"	Administers the office by coordinating all functions requiring data assembly, data correlating, reporting, document production and informing external and internal entities by directly interacting with other databases.

If you have any questions, please call us at (708) 491-7928. Ask for Gary Summers or Dr. Elie Geisler.

Code No. _____

USERS NEEDS FOR INTELLIGENT SYSTEMS

Milton H. Paul, M.D.

Phil R. Goldberg, B.S.

The Children's Memorial Hospital

Northwestern University

ICARIS - Integrated Clinical And Research Information System

DATIS - Decision And Teaching Information System

There is an increasing recognition of the role that advanced medical information processing systems can play in facilitating and improving medical diagnosis, medical-management decisions and results assessment studies. The objective of our project has been to provide for the effective use of large, heterogeneous, data-based information sets via on-line interaction to achieve more informed medical decisions and enhanced teaching.

Traditionally, in medical and surgical sub-specialty sections concerned with complex problems, physicians gather in group work/teaching sessions (Figs. 1, 2) to review a large array of diagnostic test results and clinical findings for the patient/problem under consideration. As an example, in cardiology this review may include x-ray films, electrocardiograms, cardiac catheterization studies with angiocardiograms and hemodynamic measurements, nuclear medicine studies, echocardiograms, and drawings, photographs or descriptions of operative procedures. Each result is acquired and presented via different formats, *e.g.*, x-ray films, video-tape, 35 mm cine-film, digitized-images, and printed page numerical and text reports.

We present a prototype, generic information management system, **ICARIS/DATIS**, that can provide users with an efficient, personal and departmental knowledge/experience acquisition and repository structure that can be further integrated within larger (hospital-wide) enterprise networks, and perhaps regional or universal information networks (Figs. 3, 4).

Currently, most hospital-wide information systems address primarily operational, fiscal and selected "macro"-medical departmental functions (i.e. laboratory, pharmacy, critical care, radiology) and are not directly involved with the detailed medical data management of medical and surgical subspecialties.

The objectives of **ICARIS/DATIS** are to provide this diagnosis/decision environment with the means for acquiring, manipulating, and investigating large data sets, comprising numerical, textual and imaged data, for analyzing:

[1]   a current "index" patient or problem under inquiry (Fig. 4, **Local Knowledge/Experience** - Patient X);

[2]   prior local physician, departmental or institutional experience regarding groups of patients or problems (Fig. 4, **Local Group Knowledge/Experience** - Group N); and,

[3]   outside experience, reports, and opinions retrieved from scientific journals, monographs, etc. (Fig. 4, **Universal Knowledge**).

An important objective of the **ICARIS/DATIS SYSTEM** is to provide a "knowledge - driven convergence function" (Fig. 5) by incorporating an efficient knowledge sharing paradigm, based on an object-oriented approach to information to provide rapid access to domain-specific knowledge for diagnosis, patient management decisions and problem solving.

Two key elements facilitate the acquisition and integration of this knowledge-sharing function (Fig. 6):

[1]   A centralized data server is modelled on an industry-standard, open-systems architecture, with expansive facilities for on-line storage, powerful software tools and communications abilities, and provides access to both local/collected knowledge and remotely accessible universal knowledge; and,

[2]   A portable data/knowledge collection unit (Fig. 7, comprised of a notebook computer

and hand-scanner, permits the physician (domain-specialist) to extract and file information during ordinary daily work and perusal of contemporary scientific literature or historic reports. This information, judged relevant and important by the reader, is stored for future recall using both domain-specialist (personal) and standardized (specialty) indexing terms.

The **ICARIS/DATIS** presentation format (Fig. 8) provides a comprehensive image-data-text linked screen, pertinent relational database management system search formats and comment sections.

Technologically, the broad system goals included: (1) off-the-shelf technology; (2) industry standard expansion; (3) low maintenance and easy repair; (4) speed; (5) low cost; (6) multi-user operating system; (7) robust software development tools; (8) shrink-wrapped application software; and, (9) outside communication mechanisms. Accordingly, the specific goals and minimum requirements included (Fig. 9): (1) SCSI and/or ESDI interfaces; (2) 16 Megabytes SIMM memory; (3) high resolution graphics; (4) one gigabyte of fixed disk space (with expandability for image files); (5) five million instructions per second (minimum); (6) quick floating-point performance (for statistics); and, (7) Unix™ (AT&T™ and BSD™ mechanisms).

Additionally, the present system design concept (Fig. 10) provides an overall object oriented structure with both an object oriented work environment and data structures to provide self-defining data sets and the means for functions to verify operational appropriateness and integrity.

Some examples of data that can be extracted and presented relevant to the subspecialty domain of pediatric cardiology data are illustrated as follows:

Fig. 11   Angiocardiogram (antero-posterior and lateral views) illustrating blood-flow jet at stenotic pulmonary valve orifice and extracted from one inch video-tape recording format of x-ray image-intensifier.

Fig. 12   Angiocardiogram (lateral view) illustrating two ventricular septal defects (black arrows) and subpulmonary stenosis (white arrow) in patient with transposition of the great arteries. Hand-scanned image from labelled patient record derived from 35 mm cine filming of x-ray image-intensifier.

Fig. 13   Echocardiogram (2-D) from same patient as Fig. 12 illustrating two ventricular septal defects (white arrows) and subpulmonary stenosis (black arrow). Hand-scanned image/print derived from half inch video tape of ultrasound recordings.

Fig. 14   Physician's sketch representing synthesis of morphologic findings and two operative procedures on complex congenital cardiac malformation. Hand-scanned image from patient's record.

Fig. 15   Illustration of suture placements for cardiac operation hand-scanned from contemporary cardiac surgical journal.

Fig. 16   Data-log presenting hemodynamic and blood gas measurements from operational cardiac catheterization data-base.

Fig. 17   Graph presenting statistical summary of large-scale inter-institutional results (percent survival) after arterial switch repair of transposition of great arteries. Hand-scanned from privately circulated report.

Fig. 18   An estimate of the hand-scan extraction load for one year's personal perusal of the medical literature was obtained (Fig. 18) by reviewing the 6 major (English language) cardiology and cardiac surgery journals for 1991. Approximately 1200 continuous-tone images, data-graphs and data-tables, and abstracts were considered relevant and useful, primarily for clinical application, using **ICARIS/DATIS** in a pediatric cardiology subspecialty setting. This load presents little additional time-effort during journal reading since, even with the prototype unit, initial scanning and keyword indexing without editing takes less than a minute or two per image.

A fully realized **ICARIS/DATIS** information management system - with its acquisition mechanisms, database storage, and index keying, and analysis tools - can provide a critical merging function between individual, group and universal experience/knowledge in the medical setting and create an arena wherein intelligent, informed decisions may be expeditiously reached.

Figure 1

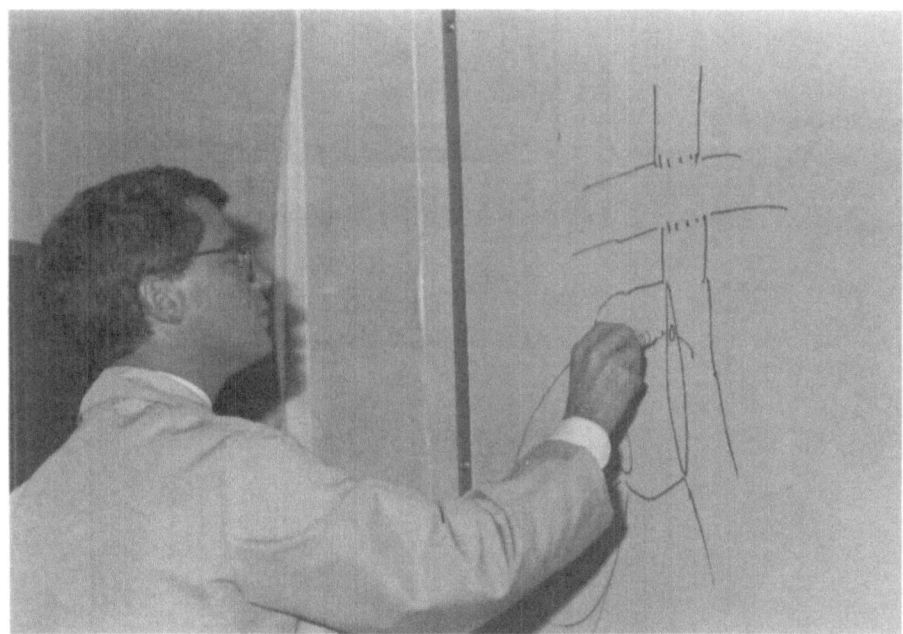

Figure 2

32

Figure 3

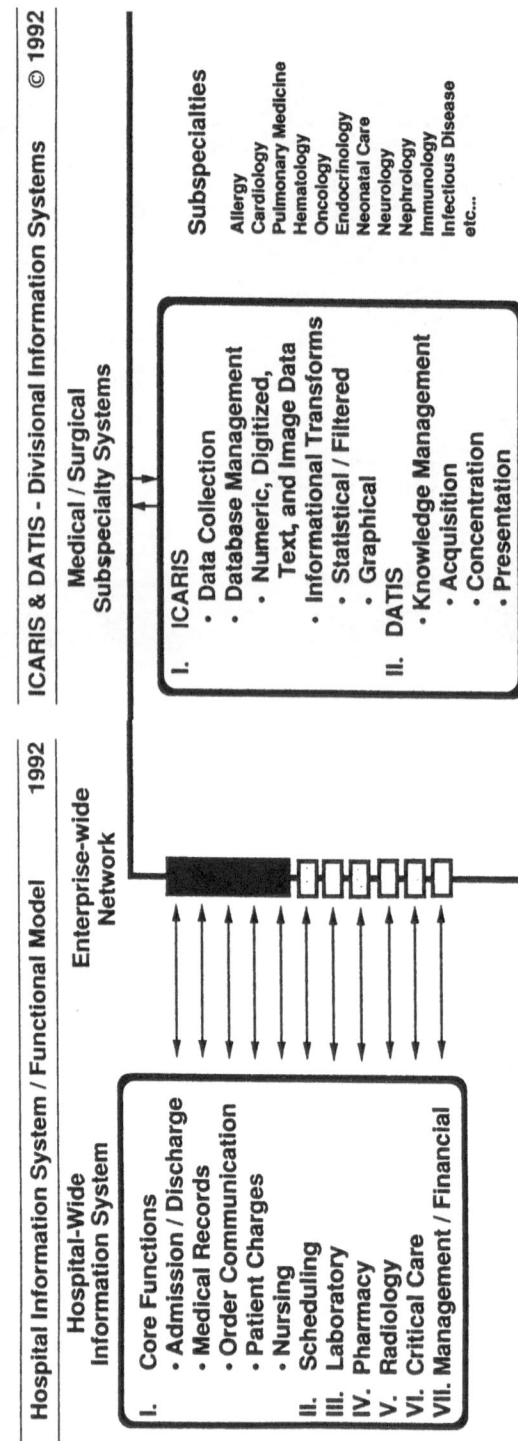

Hospital Information System / Functional Model   1992

Hospital-Wide Information System

Enterprise-wide Network

I. Core Functions
    • Admission / Discharge
    • Medical Records
    • Order Communication
    • Patient Charges
    • Nursing
II. Scheduling
III. Laboratory
IV. Pharmacy
V.   Radiology
VI. Critical Care
VII. Management / Financial

ICARIS & DATIS - Divisional Information Systems   © 1992

Medical / Surgical Subspecialty Systems

I. ICARIS
    • Data Collection
    • Database Management
      • Numeric, Digitized, Text, and Image Data
    • Informational Transforms
      • Statistical / Filtered
      • Graphical
II. DATIS
    • Knowledge Management
      • Acquisition
      • Concentration
      • Presentation

Subspecialties

Allergy
Cardiology
Pulmonary Medicine
Hematology
Oncology
Endocrinology
Neonatal Care
Neurology
Nephrology
Immunology
Infectious Disease
etc....

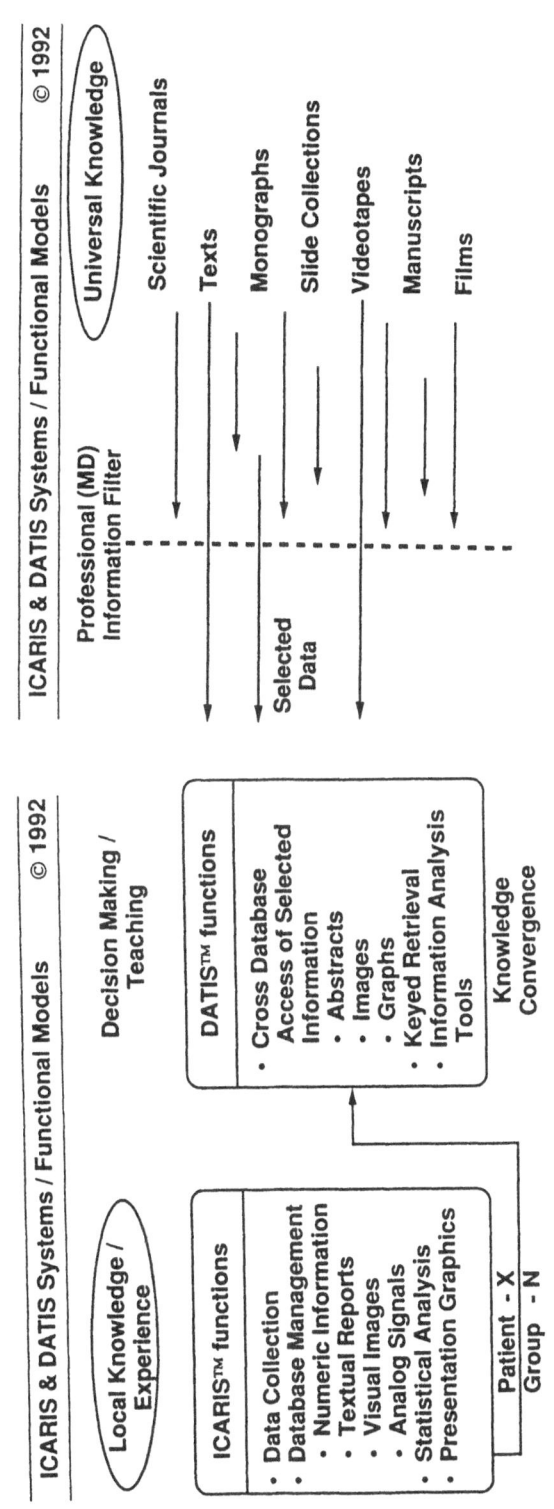

Figure 4

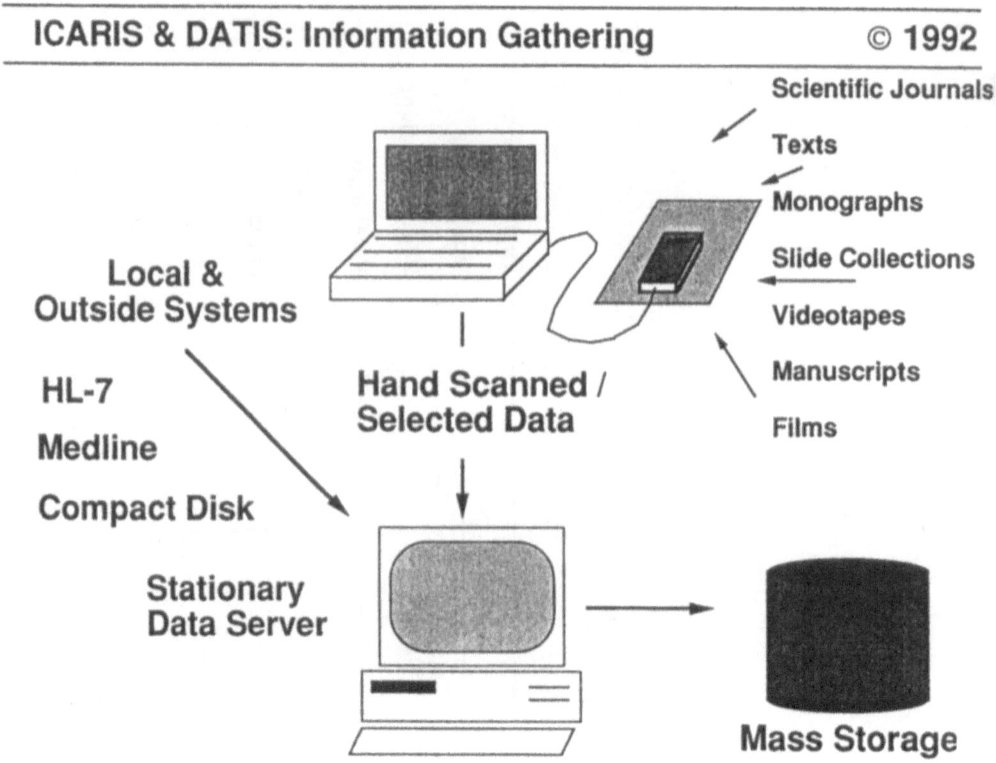

ICARIS & DATIS: Information Gathering
© 1992
Scientific Journals
Texts
Monographs
Slide Collections
Videotapes
Manuscripts
Films
Local &
Outside Systems
HL-7
Medline
Compact Disk
Hand Scanned /
Selected Data
Stationary
Data Server
Mass Storage

Figure 5

DATA, INFORMATION, KNOWLEDGE
Knowledge-Driven Convergence Function
OBJECTIVES
Index Patient [X] Local
Class of Patients [N] Local World
Area of Interest [U] World
Derived & Enhanced Knowledge
Diagnosis
Medical Decisions
Teaching

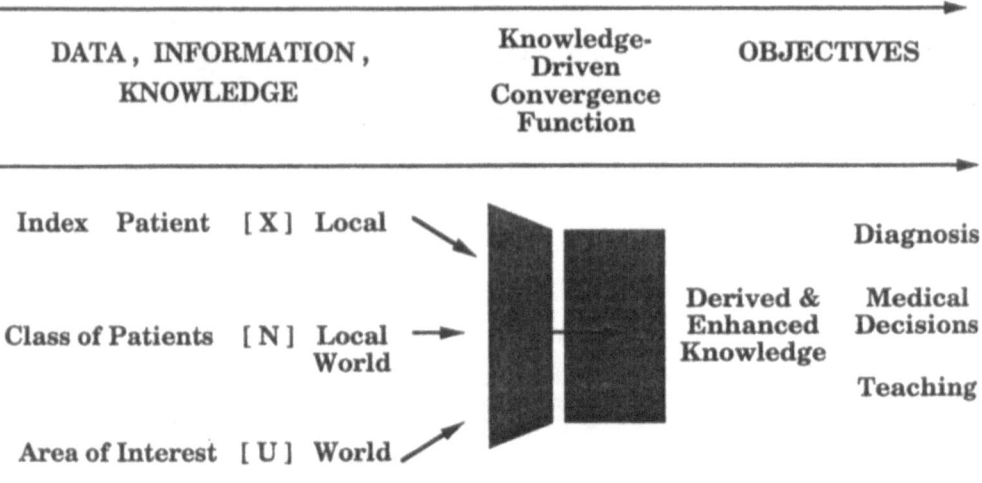

Figure 6

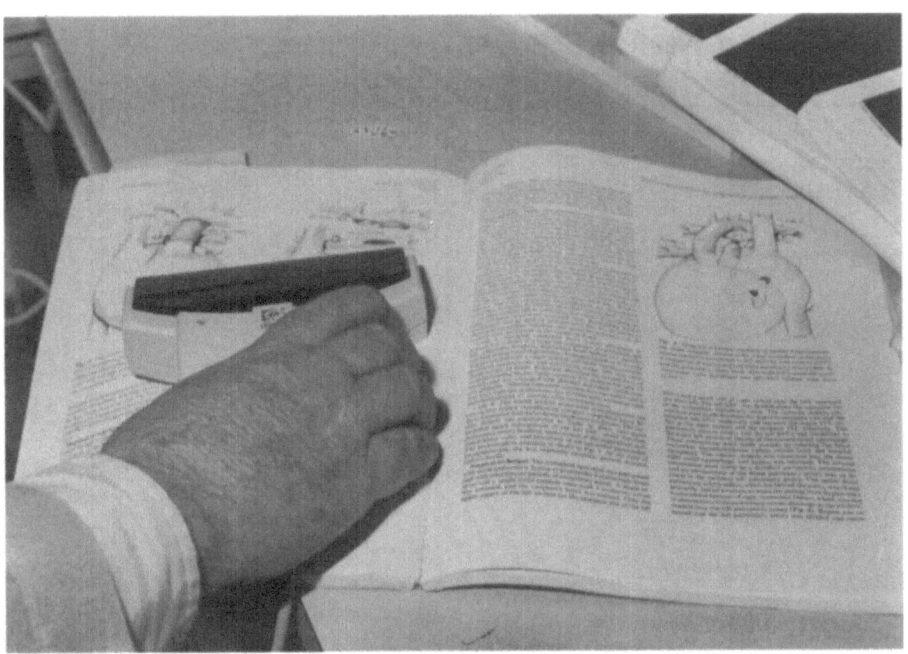

Figure 7

Screen Layout

Menu Bar	
Image	Numeric / Textual Data (Linked to Image)
Legend	Commands
Image Comments - Source - Value / Quality - Enterer	Formal Search Keywords / Personal Search Keywords

Figure 8

ICARIS: Physical System Configuration © 1992

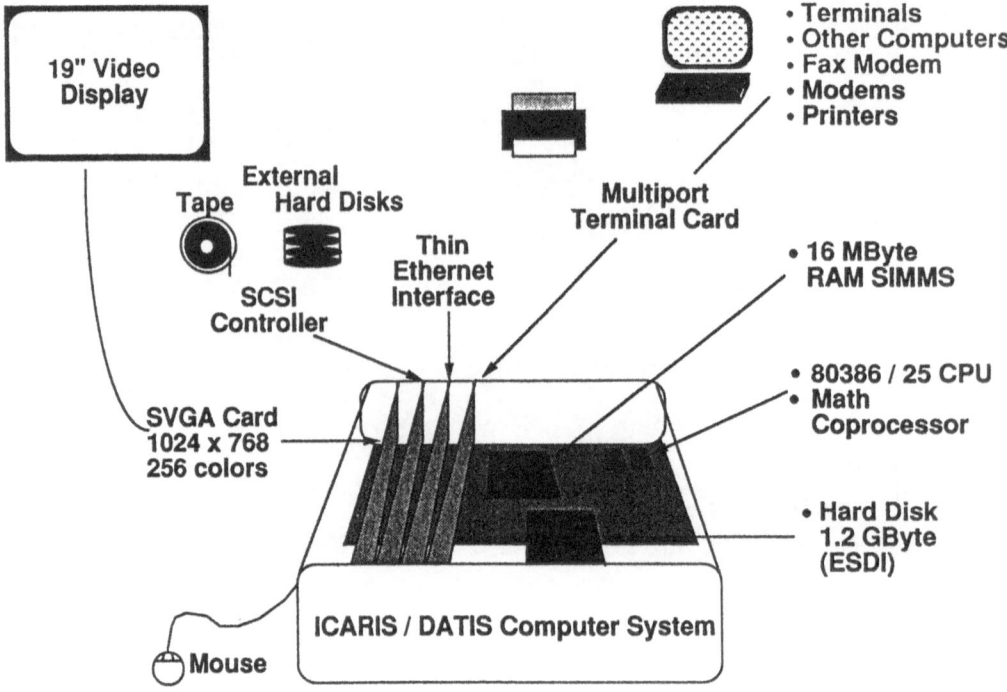

Figure 9

Object Oriented Environment

Data History, Encapsulation,
and Conversion Layer

DATIS LAYER
• Cross Database Searching
• Linking of Data Together

Basic Applications Layer

ICARIS CORE
• Communications
• Data Storage &
  Retrieval

Unix™ Operating System

Figure 10

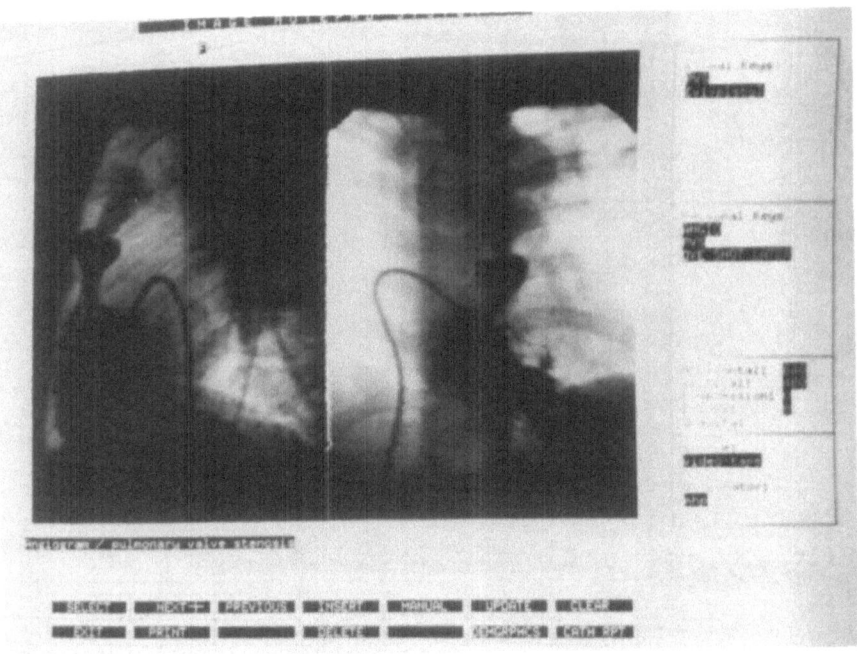

Figure 11

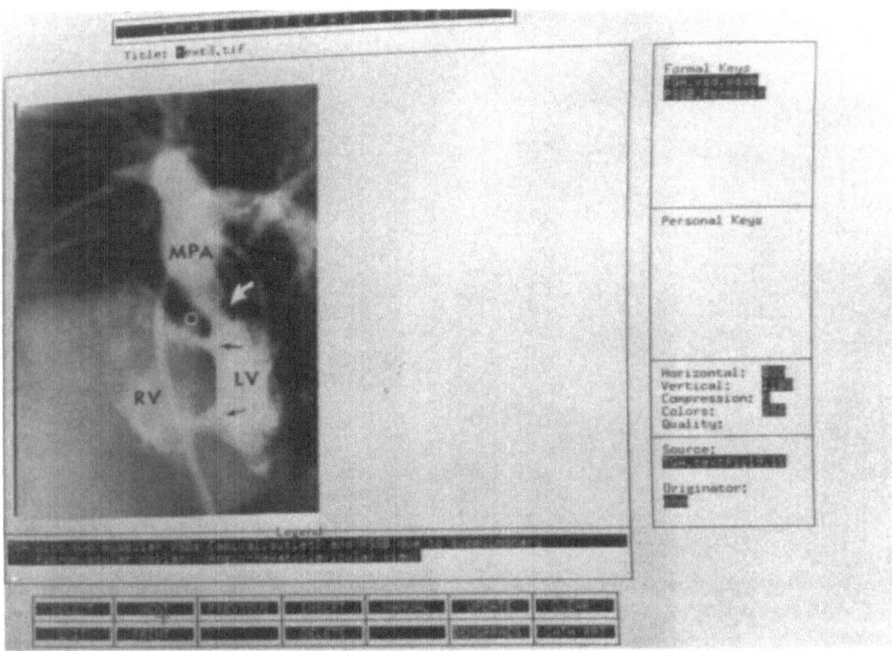

Figure 12

38

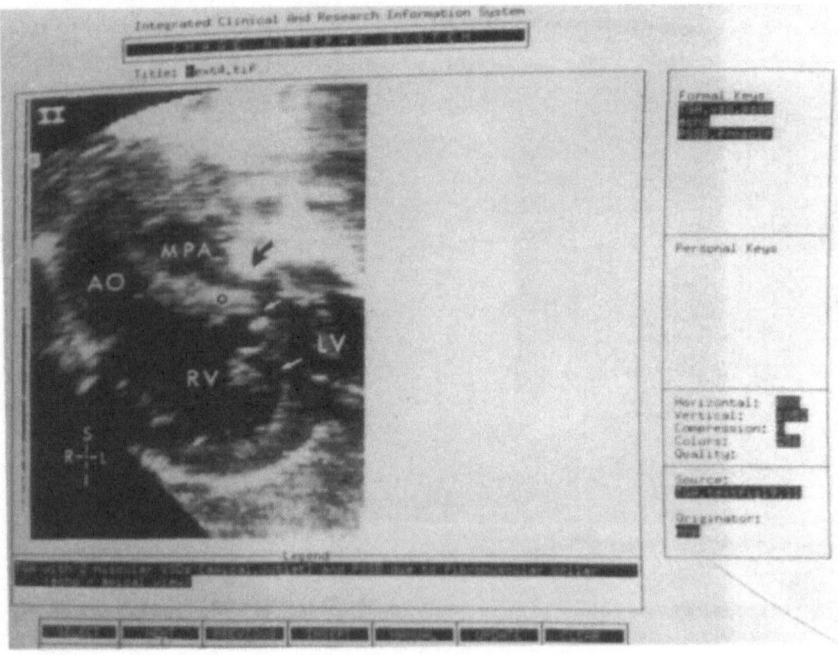

Figure 13

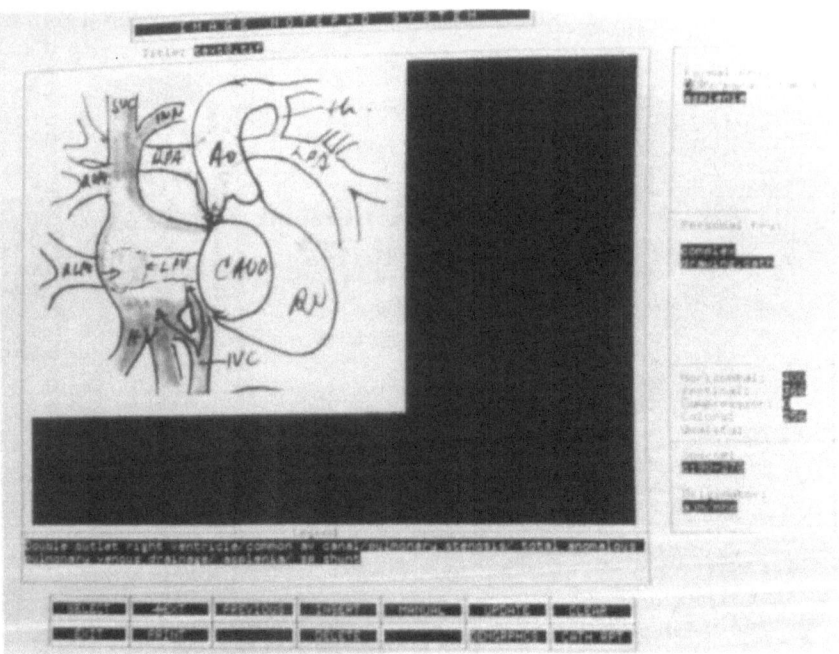

Figure 14

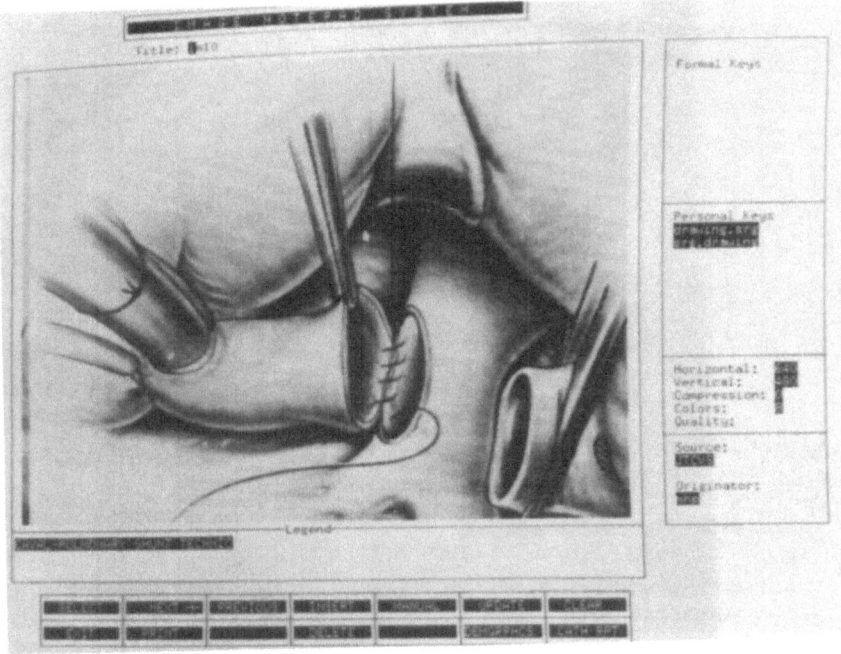

Figure 15

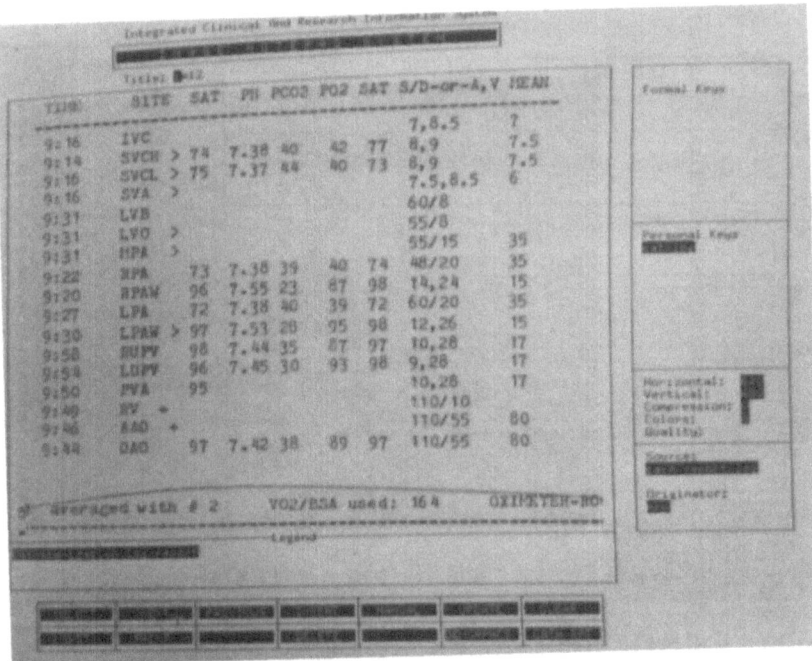

Figure 16

40

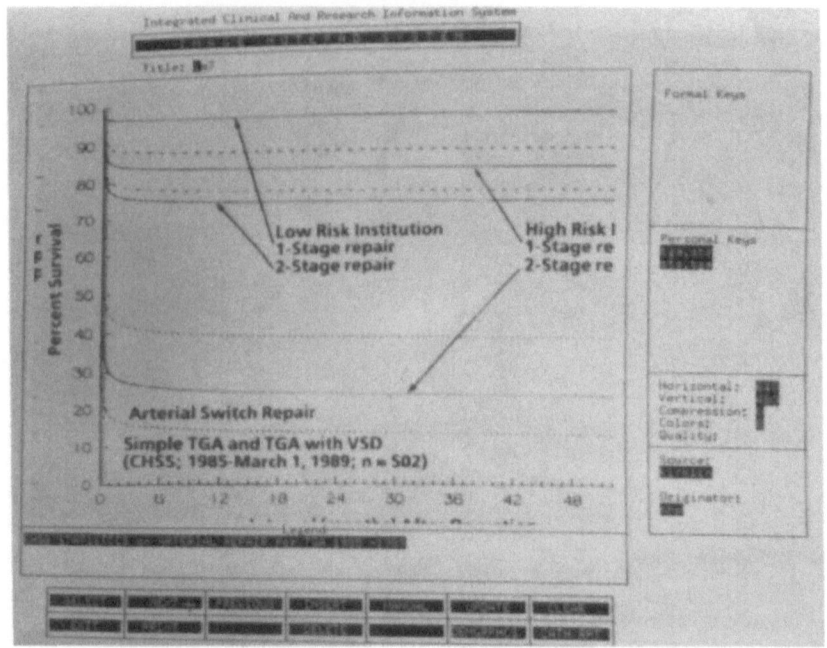

Figure 17

HAND-SCAN LOAD

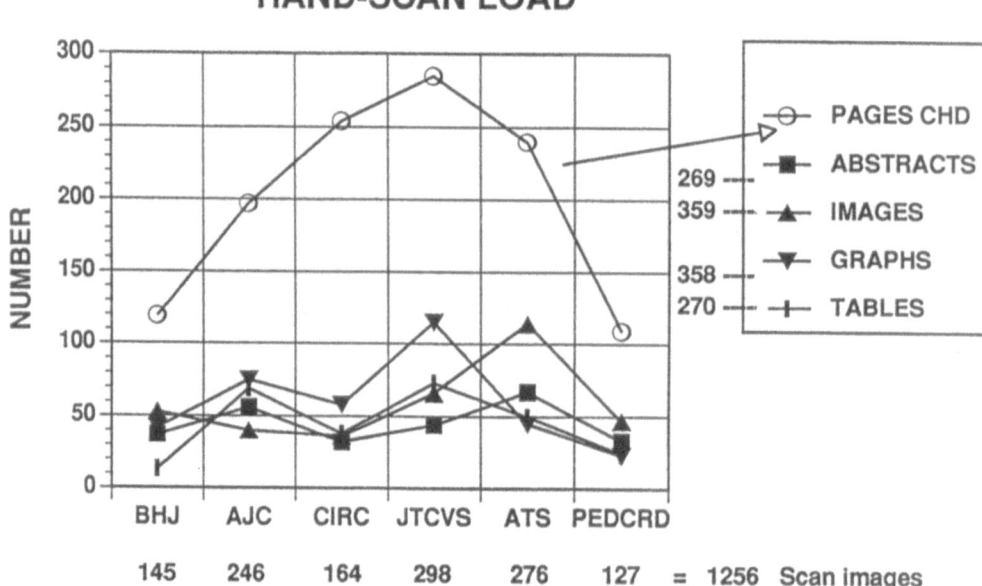

Figure 18

Opportunities For Research
To Meet Customer Needs

Gerald M. Hoffman
Northwestern University

Abstract

Northwestern University's research on User Needs for Intelligent Systems (UNIS) has identified more than twenty-six different intelligent systems which managers and professionals specified as potentially helpful in their jobs. Others are appearing as the research continues. Although these systems are very different externally, internally they can be driven by a relatively small number of different kinds of programs, or engines. An analysis of the functional requirements placed on these engines shows several areas where further research could yield commercially important results. These include

The syntactic/semantic structure of natural language.
The syntactic/semantic structure of information in forms other than language: images, sound, video, and additional modes not yet invented.
Effective and efficient storage and retrieval methods for extremely large heterogeneous data bases.

Background

Northwestern University's Center for Information and Telecommunication Technology is engaged in a multi-phase research project on Users' Needs for Intelligent Systems (UNIS). This paper reports some results from the first phase of that project, which included semi-structured interviews with nearly one hundred professionals in fields such as medicine, law, and architecture.

Definitions

These definitions have been adopted for the purposes of this paper.
Application
A information system immediately usable for a business purpose, (e. g., an accounting system built using the dBase IV product).

Machine
> A system which can be used to build applications, (e. g. dBase IV itself).

Engine
> The core programs within a machine which give it its unique capabilities (e. g. the storage algorithms within the dBase IV system).

An intelligent system is a particular type of application, defined below.

A Taxonomy of Information Systems

Use of the term "intelligent systems" immediately raises the question of how they are different from other applications. The taxonomy described below relates intelligent systems to the larger world of information systems.

The uses of information systems as business tools (i. e., applications) can be divided into these categories:

> Data processing systems -- information systems that support business processes which are so well understood the they can be described completely by algorithms.

> > Payroll, accounting, and inventory control are examples.

> Decision support systems -- information systems that deal with business processes which are not completely modelled by algorithms, either because it is impossible to do so, or because it is very burdensome. Some parts of the process are modelled by algorithms, and the balance are handled by human beings in close conjunction with the computer-based parts.

> > The use of spreadsheets to analyze business alternatives is a good example. In many cases it is conceptually possible to use mathematical programming models to achieve the same results, but the complexities of the models plus their inherent theoretical limitations make a machine-human interaction system the preferred solution.

> Expert systems -- systems that deal with problems not understood well enough to be modelled by algorithms. The analysts attempt to model the ways in which human beings solve problems, rather than attempting to model the processes which create the problems.

Published examples of expert systems now number in the thousands, including such diverse areas as diagnosis of human disease, extending commercial credit by banks, repairing railroad locomotives, and designing information systems.

Intelligent systems -- systems which model both the algorithmically accessible parts (as do decision support systems) and the non algorithmically accessible parts (as do expert systems) *without requiring the intimate man-machine interactions during use which are characteristic of decision support systems.*

The requirement of decision support systems for intimate man-machine interaction is anathema to managers and to many other professionals. Observations that managers and other professionals such as physicians prefer not to work with keyboards are so ubiquitous that they do not require citation here. Intelligent systems will fail in the marketplace, whatever their capabilities, in direct proportion to the amount of human interaction required for their use.

Intelligent systems will probably be built using techniques from all types of information systems: conventional data processing, decision support, operations research, and expert systems and other artificial intelligence methods. The key to success will not be the techniques employed: it will be the definition and specification of system capabilities which are both useful and acceptable to those who buy them.

Intelligent Systems Wanted By Users

Over twenty six specific needs for intelligent systems were identified by managers and professionals in interviews during the first phase of the Northwestern University research project. A representative subset of the systems is described briefly, each as specified by a particular potential user. The duplications and redundancies are analyzed in subsequent sections of this paper.

1. Disaster Backup System
   A system which automatically prepares for possible failure of a computer system, and which takes action without human intervention

in the event of failure.

2. Predicting Future Evolution of Law System

A system to predict the future evolution of law and regulation in a specific area. The system would analyze new legislation, court decisions, and pending legislation to determine whether regulation was likely to become more or less stringent.

3. Parsing Legal Documents System

A system to help assure that a complex legal document is internally consistent, by analyzing the logical consequences of various inclusions and exclusions.

4. Identifying Characteristics of a Deal System (Law)

A system used by an attorney during the first meeting with a client to identify all of the information needed for the attorney to do his/her work. For example, in preparing a will, the attorney needs to know all family relationships, the nature and location of all assets, etc.

5. Completeness Assurance System (Law)

A system to help with the end of a legal process, identifying and tracking all of the documents and actions required complete a transaction.

6. Predicting Opponents' Responses System (Law)

A system to trace and analyze previous actions of opposing principals and attorneys, and to predict their response to various possible actions.

7. External Consistency Assurance System (Architecture)

A system to assure that the construction specifications for a new building meet all applicable regulations of governmental bodies and all industry standards, while also meeting the client's requirements. These regulations and standards are voluminous, are often revised, and are sometimes inconsistent with one another.

8. Overall Systems Integration System

A system to provide easy access for everyone in an organization to all information in every computer system in the organization.

9. Conflict Resolution System

A system to help reconcile the diverse opinions of various specialists in an organization, such as disagreements about a new product among financial specialists, marketers, and engineers.

10. Hopes and Dreams Extraction System

A system to help people go beyond the everyday in seeking solutions to problems. Such a system would have been extremely useful in doing the research which is the topic of this paper.

11. Automatic Meeting Reporting System

A system to record the discussions at a meeting, analyze their content, rearrange them into logical and action oriented form, and distribute the results as required.

12. Focused Information Collection System

A system to retrieve information from both internal and external sources according to preselected criteria, analyze the content of the retrieved material, and filter out redundant material.

13. Synergistic Information Retrieval System

A retrieval system which analyzes the content of retrieved material and automatically retrieves related material.

14. Discovery Standardization System (Law)

A system which assures that all relevant questions are asked of each witness during discovery, a pretrial legal process in which each party has the opportunity to question the other party's witnesses so that there will be no surprise testimony during the trial itself.

15. International Transaction System

A system to assure that an international transaction complies with the laws of all countries involved, by identifying all relevant laws and analyzing their content to detect requirements and inconsistences.

16. Audio to Hard Copy System

A system to convert speech into printed text.

17. Money Management System for Professional Offices

A system to manage the financial and business affairs of a professional practice such as law or medicine.

18. On-line Information Extraction System (Law and Medicine)

A voice driven system to extract reference information from technical databases, both local and remote.

19. Mentoring and Apprentice Training System (Law)
A system to help newly graduated attorneys understand law firm policies and procedures, and learn some basic facts not taught in law schools. ("How to find the courthouse" is an often quoted example.)

20. Generalized Office Management System (Law, Other)
A system to support the management of a law practice, including both financial and operational information.

Classification by Business Opportunity

The needs expressed by potential customers for the machines listed above present various kinds of business opportunities.

Marketing opportunities, because the applications described already exist:
Systems 1, 17, and 20 (for many industries) exist as products currently on the market.
Systems 9 can be an application of existing groupware.

Development opportunities, because the machines or the necessary engines exist but the applications do not.
Systems 4 and 5 can be built using a word processor.
Systems 7, 14, and 19 can be built using expert system shells.
System 8 can be built using various existing techniques.
System 16 is a straightforward application of speech recognition.
(The need for a general speech recognition engine is well known and is currently the object of many research efforts. Since our goal is to identify new needs for research, we assume for the purposes of this paper that such an engine will be available when needed.)

Research opportunities, because engines do not exist but appear to be feasible.
Systems 2, 3, 6, 10, 11, 12, 13, 15, 18.

Common Engines

Analysis of the systems identified as research opportunities shows that five basic engines would meet the needs of all of them -- as well as the needs of other intelligent systems not mentioned in this paper.

Extensive Search and Filter Engine

This engine has (1) full key word and Boolean logic search capabilities, (2) automatic access to multiple databases, both public and private, and (3) sophisticated filters which minimize the retrieval and display of redundant or irrelevant information.

> An ordinary retrieval by key word might yield fifteen or twenty news articles about a corporate earnings report. The filter could be set to display only the longest, the shortest, and citations of the others.

Free-form Interactive Brainstorming Engine

This engine extends the concept of unfettered exchange of ideas between people ("brainstorming") to the exchange of ideas between people and databases.

> A participant's use of the phrase "quality in information systems" might cause display of a video of a speech on the topic, as well as written material relating to the more general issue of quality concepts for staff services.

Semantic Analysis and Comparison Engine

This engine performs semantic analysis and content analysis of both text data and other forms of data (images, etc.), and compares separate data streams for logical consistency.

> Attorneys in a real estate transaction representing buyer, seller, and lender might each prepare a document proposing contractual terms. The engine would analyze the three proposals, and perhaps a land survey as well, to detect areas of agreement and disagreement.

Synergistic Concept Generator

Data retrieved by standard Boolean search is analyzed for content to identify concepts related to the original search criteria. These are used as the basis for further retrievals. The process may be iterated as often as desired.

> A retrieval based on the key words "marketing" and "European community" might find many articles which also contain references to "ISO 6000". A further retrieval based on "ISO 6000" might point toward the concept of Total Quality Management, and so on.

Verbal Projection Engine

This engine makes predictions by analyzing time series of words in ways analogous to statistical predictions based on time series of numbers.

> A set of references to the development of speech recognition systems might contain the terms "distant", "progress", "setback", "nearly", and "prototype", found in the time sequence shown. The engine would predict the nature and timing of the next phase of development, together with an appropriate confidence level.

Fitting Engines to the Machines

The list below shows which of the new engines described in the previous section are needed by each system identified as a research opportunity.

2. Predicting Future Evolution of Law System
   Extensive Search and Filter Engine
   Semantic Analysis and Comparison Engine
   Verbal Projection Engine

3. Parsing Legal Documents System
   Semantic Analysis and Comparison Engine

6. Predicting Opponents' Responses System (Law)
   Extensive Search and Filter Engine
   Semantic Analysis and Comparison Engine

10. Hopes and Dreams Extraction Machine
    Free-form Interactive Brainstorming Engine
    Semantic Analysis and Comparison Engine
    Verbal Projection Engine
    Synergistic Concept Generator

11. Automatic Meeting Reporting System
    Semantic Analysis and Comparison Engine
    Synergistic Concept Generator

12. Focused Information Collection System
    Extensive Search and Filter Engine
    Semantic Analysis and Comparison Engine

13. Synergistic Information Retrieval System
    Extensive Search and Filter Engine
    Semantic Analysis and Comparison Engine
    Synergistic Concept Generator

15. International Transaction System
    Extensive Search and Filter Engine
    Semantic Analysis and Comparison Engine
    Synergistic Concept Generator

18. On-line Information Extraction System (Law and Medicine)
    Extensive Search and Filter Engine
    Semantic Analysis and Comparison Engine

General Research Issues

Building the engines described in this paper will require new insights in several areas of basic knowledge.

The syntactic/semantic structure of natural language.

> As knowledge work becomes the dominant form of productive activity, the importance of information in the form of text increases dramatically. The ability to analyze, manipulate, and synthesize text by computer based methods will be essential to every organization.

The syntactic/semantic structure of information in forms other than language: images, sound, video, and additional modes not yet invented.

Images are currently an important information acquisition and storage medium in fields as diverse as petroleum exploration, insurance claim processing, and forensics. Multimedia computer systems are adding sound and video to the palette of the information system designer and user.

One requirement for effective use is the ability to translate information from one mode to another: from image to text, or text to sound. To accomplish this requires knowledge about both structure (syntax) and content (semantics) of all modes.

Content analysis of these forms of information is at least an order of magnitude more difficult than content analysis of text. It is also probably an order of magnitude more important.

Effective and efficient storage and retrieval methods for extremely large heterogeneous data bases.

The addition of sound, image, and video data to the already massive amounts of text data which must be stored and retrieved will create new challenges for computer science, even with increased understanding of syntactic/semantic structures.

The research results also raise some issues about the people who are the potential users of this technology.

Why don't they know about things which are readily available which meet their needs?

What makes people want to buy and use this kind of technology? What interferes?

Acknowledgements

The other members of the UNIS team provided valuable information and insights during this research. The research was supported by Siemens AG.

INTELLIGENT SYSTEMS: NEW PRODUCT DEVELOPMENT

Nagaraja R. Srivatsan
McCormick School of Engineering and Applied Science
Northwestern University

Abstract

Intelligent systems (IS) are defined as the next generation tools and products which can provide intelligent assistance to users in their day to day professional activity. A methodology is provided to identify and develop intelligent systems in the future. A market segmentation strategy is proposed to market such intelligent systems. Finally, the methodology and segmentation strategy is applied to the current Expert System industry.

1 Introduction

Intelligent systems (IS) are the next generation tools and products which can provide intelligent assistance to users in their day-to-day professional activity. IS are a set of tools which will aid professionals in performing their jobs better by providing alternate decision paths, "playing field for ideas," effective communication and retrieval with different sources of information, and learning mechanisms for professionals to educate themselves on different topics of interest related to their profession. Intelligent systems can be defined as the next generation of artificial intelligence products. AI researchers give the following specifications, including a variety of features which make a system intelligent [10].

1. The systems knowledge and understanding should be coherent, *i.e.*, what is known by one module should be known by the other modules.

2. The data or information should be accessible to the system, regardless of the mix of structures used to maintain the data, the physical separation of data in the different files and the indexing mechanisms of these files.

52

3.   The knowledge and understanding codified in the system should be readily accessible to the end user or the end system, the manner of the dialogue should be uniform, and the system should understand a wide variety of questions posed in natural language format.

4.   Intelligent submodules of the overall system should be able to initiate appropriate actions on their own; this implies a high level of software integration.

5.   A flexible procedure to modify the expertise of the system should be available and possible inconsistencies that arise due to the modification should be identified and resolved.

6.   The system will do some things that may not have been formally anticipated by the creator, but which an intelligent person would do in similar circumstances. The system has the ability to solve new and different problems [21].

7.   Intelligent systems must execute rapidly, integrate with existing systems and applications, and access data from a variety of environments in a manner in which the physical location of the data is transparent to the user.

Current artificial intelligence products like expert systems, vision systems and natural language processing systems do not integrate well with each other and fall short of the definition of intelligent systems as stated earlier. Artificial intelligence research has been trying to develop intelligent systems as defined above for more than three decades. From the researchers' initial euphoria of developing intelligent machines which can think, artificial intelligence research is currently taking a more pragmatic approach to making products which can solve specific problems in a given domain of knowledge [13].

Artificial intelligence products historically have been developed more from a research and a technological perspective than from the users' point of view [38]. The problem of new product identification for intelligent systems should be approached from a user-needs perspective rather than primarily from a technological perspective. New product ideas can arise from various sources like patents and inventions, competition, user needs, user solutions, technology, engineering, management and employees [43]. Designing new products by identifying user needs tends to be successful [43]. User needs provide information that can help the company position itself in a better manner in the market and identify opportunities to develop new products. Typical market research questions to identify users needs tend to fall into three main categories:

1.	Questions about product

2.	Questions about needs or difficulties and

3.	Questions about novel solutions the user has found for their problems [42].

New product identification for intelligent systems tends to be difficult, using conventional market research questions. Intelligent systems do not exist, thus making it difficult for the user to answer questions on the product or the use of the product. A combination of conventional market research questions and cognitive probing questions were used by the User's Needs for Intelligent Systems (UNIS) team to obtain users' needs for IS. The cognitive probing technique was used to probe the thought processes involved in carrying out the professionals' daily activities and to identify "core" intellectual activities required by the professionals to perform their occupation effectively.

A methodology is described in section 11 for identifying new products for intelligent systems from a user-needs perspective. A set of common components were extracted from the potential new products identified [32]. The current state of technology and possible research direction to be taken to realize the new products will be discussed. Segments of users with similar characteristics are likely to exhibit similar product usage behavior; these segments will be identified. Finally, the methodology and the segmentation strategy will be applied to the current Expert System industry.

2   Definitions

1.	**Knowledge**: Knowledge is defined as facts or ideas, acquired by study, investigation, observation or experience.

2.	**Domain**: Domain is defined as a region distinctively marked by some characteristics. In the following sections, domain refers to regions in a particular subject matter that have clear and well defined characteristics that demarcate one region from other regions. For example, real estate law is a domain. The domain addresses real estate law issues and is a part of the subject matter, law. Cardiology is a domain related to the study of heart, its action and diseases and is a part of the subject related to human health.

3.	**Data**: Data consist of factual information (as measurements or statistics) used as a basis for reasoning, discussion or calculation. Data contain information but often not in the form and format users can directly use to make decisions. In the domain of marketing, point-of-sale data are available from bar code scanners installed in different retail outlets.

4. **Information**: Information is defined as the communication or reception of knowledge. Information is referred to as a quantity that is provided by the system to users in a form and format users can use to make decisions. An example in the domain of marketing would be providing strategic marketing personnel with what the mix of a new product should be, based on point-of-sale data.

3 Intelligent Systems: Past and Present

From its days of conceptualization, artificial intelligence research has been trying to define intelligence and intelligent systems. The different definitions of "intelligence" have led artificial intelligence research in many directions. Defining intelligence as being smart, smart game playing programs like Deep Thought for chess and smart checker programs were built. Defining intelligence as ability to plan, planning programs like General Problem Solver (GPS) were developed by Allen Newell and Herbert Simon [28]. Defining intelligence as a repository of an expert's knowledge, Expert Systems were developed for different professions. Defining intelligence as mechanisms of learning, Neural Network applications have been developed. Expert Systems have become commercially synonymous with artificial intelligence products and there has been a proliferation of Expert Systems in the market.

Expert Systems are specialized Artificial Intelligence (AI) programs which embody knowledge which has been extracted from a human practitioner and which has been organized and computerized. This system would then simulate the deductive logic of the human expert. Expert Systems are, however, unable to learn from experience, and hence their knowledge base must be updated through heuristic human intervention [38].

The early Expert Systems included DENDRAL [2], MYCIN [4], Internist MAC-SYMA [23], HEARSAY I and II [30] and PROSPECTOR [9]. These systems were primarily undertaken for exploring the technologies of Expert Systems. In the late 1970's an experiment at Stanford resulted in abstracting EMYCIN from the MYCIN "parent" [4]. The resulting shell was used to implement PUFF, an Expert System for interpreting pulmonary functions. This led to the development of Expert System shells, the tools to develop Expert Systems. The growth of Expert Systems led many companies to offer programs to train personnel to be knowledge engineers, *i.e.* Engineers who build Expert Systems. Expert Systems were a high growth industry in the early 80's. There were Expert System products in nearly every industry. In medicine, METADENDRAL [2]; in Accounting, PLATINUM LABEL (IntelliSource, Inc.); for taxation systems, TAX ADVISOR (University of Illinois); in insurance, USER. There were many Expert System shells developed in the 80's like ART (Automated Reasoning Tool) by Inference Corporation. KEE (Knowledge Engineering Environment) by Intellicorp, KES (Knowledge Engineering System) by Software Architecture and Engineering Inc., Knowledge craft by Carnegie Group, GURU by Micro Data Base Systems Inc. Expert Systems are one class of the many artificial intelligence products available that symbolize the artificial intelligence industry today. Most people interchange the terms artificial intelligence and Expert Systems. Until many more artificial intelligence products become commercially viable, Expert Systems will reflect the state of artificial intelligence.

Current Expert Systems development requires a large amount of software engineering effort and a large amount of time spent in knowledge-capturing. Knowledge-capturing is the process whereby expert knowledge in a particular field or domain is captured in an application. Current day Expert Systems are limited in their ability to infer and adapt, and function primarily by storage of information [11]. In the late 80's Expert System growth has tapered off due to not meeting the expectations of many users [27].

A continuing goal of AI research is to teach computers to understand spoken language. Speech recognition research can distinguish three levels of speech understanding currently [13]: *Isolated word recognition.* The above technique is used to implement solutions for the physically impaired, programming applications in which vocabulary is restricted and processing of data in difficult situations (recording of part numbers in a poorly lit warehouse for instance). *Continuous speech understanding in a limited domain.* Products include HEARSAY-II (Carnegie Mellon) [35] and HWIM (from Bolt Beranek Newman) which provide understanding of up to 5000 words in a single domain by a single or limited number of speakers. *Continuous speech understanding - general case.* IBM is currently developing the Intelligent Typewriter which currently has about 97% recognition for IBM's word office task vocabulary spoken by any person.

Expert Systems continue to be one of the few available AI products which are marketed commercially. Neural Networks is another burgeoning industry and has grown by more than one hundred percent over the last two years [27]. Neural Networks are used for manufacturing and business applications like process and quality control, product design and analysis, and data-trend analysis and forecasting. A few of the Neural Network vendors are HNC Inc., Integrated Inference Machines, INTEL CORP., NESTOR, NEURALWARE and WARD systems. Researchers are currently working towards designing systems which can combine the features of Expert Systems and Neural Networks [22].

4   Research Methodology

Identification of new products by merely questioning professionals on their needs for intelligent systems may lead to skewed results. Professionals may not possess the intuitive ability to visualize aids that would increase their productivity. The interviewing process to be described later was used to identify users' needs. Twenty-five in-depth interviews were conducted with professionals in eight different organizations in the areas of law, health care, insurance, architecture, and R&D/Engineering in both the private and public sectors [32], [34].

Further, twelve strategic decision makers in the Expert System industry were interviewed. They were questioned about their firms' plans for the future and their perception of user needs for the future. They provided answers from both their strategic point of view as well as the experiences of their current customers.

Interviews were conducted with MIS managers of two companies for their input on User Needs. This group was unique, for they were a part of the user community as well as internal "vendors" to their user community.

The above groups provided information on potential intelligent systems. Most of the potential IS might be developed for a broad spectrum of users but some were specific to the professionals interviewed. Similar components among potential IS were extracted.

These common components form research/product problems which have to be addressed in order to develop potential IS identified from User interviews. The interviews also provided data on user preferences on interfaces to IS.

"Components" are building blocks of future intelligent systems. There are different user profiles which require different components or features in their product.

Although most of the machines (IS) and components suggested do not yet exist in the form proposed, the methodology described in section 11 can be used as guidelines to identify and develop new products for intelligent support to professionals.

5 Perception of User's Needs by Vendors

Twelve interviews were conducted with strategic marketing and sales personnel in the Expert System and Neural Network industry.

The current Expert System and Neural Network industry is a fore-runner of the future intelligent systems industry. The vendors are in close contact with users and have information on the acceptance of Expert Systems/Neural Network products. The experience gained by the vendors in developing and supporting these products provides valuable inputs into the development process of future IS.

The vendors were interviewed over the telephone. The interview took an average of 30 minutes. The following are the propositions from vendors about User's needs and their perception of the strategic direction the Expert System industry is taking.

1. *Artificial intelligence features are going to become a part of standard software.* Many Expert System shell vendors have found an increasing demand from users to integrate their products with standard software like spreadsheets, databases and wordprocessors. Future spreadsheets, wordprocessors and databases will have some degree of intelligence about a particular domain (specialized area of knowledge - for example, real estate law, heart surgery, automobile repair, etc.) and will help them intelligently process the data and store the summary of the data as useful information. In the future, standard software -- for example, spreadsheets, can provide a set of domains the user can choose, depending on the users' needs.

2. *Users are demanding that future intelligent systems should seamlessly integrate with current information systems.* Future intelligent systems should derive a part of their intelligence from the data which current day information systems possess. The intelligent systems should be able to access data seamlessly and derive inferences automatically.

3. *The acceptance of Intelligent systems depends on the ability of the system to interact with the user in a natural manner.* The users are demanding better and easier mechanisms to interact with intelligent systems. The current trend for user interface is towards "click and control" mechanisms. The users prefer interfaces that integrate in a natural manner, like voice and handwriting. Users prefer mechanisms which reduce the amount of time required to learn how to use the product. The user prefers input data in text, graphics and voice forms, and similarly to receive outputs in text, graphics and voice forms.

4. *Some Expert System/Neural Network vendors find the upper management as a major bottleneck in the acceptance of intelligent systems.* The vendors feel that upper management does not understand technology as well as people in the lower echelon. Those who understand their own needs seem to understand the potential for intelligent products, but upper management may be wary of investing in intelligent products before these systems have a successful track record.

5. *Users want the flexibility to tailor the intelligent system, depending on their skill level (expertise).* Users' skill level in a particular domain keeps changing. Users relatively new to an intelligent system require more help on the use of it. After the users becomes familiar with the system, they desire faster and more intuitive help.

6. *The software development cycles will change and provide the user with more flexible products.* Expert System shell vendors perceive a change in software development cycles. In current software development processes, 30 percent of the time is spent in development and 70 percent is spent in maintenance and training [27]. With new technology like object oriented programming, software reusability, and code generation programs, the time needed for maintenance of software systems will decrease and more time will be available for software development.

7. *Intelligent products of the future will be in areas where conventional software fails.* The customer will be demanding products in the areas of network management, planning, scheduling, quality control and inventory management. These areas have been difficult to approach with conventional software methods [27].

6 Perception of User's Needs by MIS Personnel

Two interviews were conducted with MIS personnel. A combination of the users' needs questionnaire and Vendors' questionnaire were used to interview MIS personnel.

1. *Users require appropriate interfaces to intelligent systems based on their expertise in the functional area of the intelligent system.* This proposition arises out of situations where MIS departments have to purchase different application programs depending on the expertise of the user in the functional area as well as their expertise in using the application program. An example is a "draw program." Depending on the users' expertise as an artist or a novice, the "draw program" interface differs. A novice would require the most primitive blocks to create a picture, while an artist will be more comfortable with abstract blocks to create a picture.

2. *Users want tools which convert volumes of "data" into "information."* MIS professionals have requests from users to convert the volume of data the MIS department collects to information which can aid decision making processes.

3. *Users want the profile creation process of intelligent systems to be based on both user's estimate of their profile and the system's estimate of their profile.* Users possess different profiles. They may be novices in certain aspects of their functional area and experts in others. Profiles are determined in order to help the intelligent system provide better functions to users.

7 Summary of User Needs

The following are some features potential IS need to possess in order to ease barriers to acceptance. This list was extracted from the user/vendor interviews.

1. *Inference Explanation*

   Intelligent systems need to have mechanisms that provide inference explanations. The explanations help the user better understand the IS. Professionals are wary of IS, especially those intended to aid in their professional activity, if they do not have a thorough understanding of the mechanisms by which the intelligent system arrives at its results [27].

2. *Profile changes option*

User profiles change with time and experience. Therefore Intelligent systems should provide options for users to change their user profile in real time.

3. *User interface for different profiles*

User interface for IS should be natural and easy to use. IS can accept input and provide output in different forms, such as graphics, text, voice and existing standard software packages. User interfaces should be designed to be adaptable and flexible to user's profile changes. A user interface should help the user spend less time learning how to use it. It should interact as naturally with the user's working style as possible.

4. *Flexibility to Update Knowledge*

The success of any intelligent system depends on the adaptability the intelligent system possesses in accepting new knowledge. Intelligent systems should be able to acquire knowledge in real time [22].

5. *System limitation information*

IS limitations should be clearly explained and should be accessible as a part of the intelligent system in real time. Vendors interviewed feel that overselling intelligent systems raise user expectation and users finding limitations in the IS leads to rejection of intelligent systems. IS that are totally accurate and reliable are difficult to develop [13]. It is desirable for IS to provide professionals with explanations and the reliability factors on inferences made. The user's understanding of the functions and limitations of intelligent systems leads to a more realistic use of the intelligent systems [27].

6. *Role of IS*

Professionals perceive future IS as threats to their jobs. Vendors of IS need to identify the role these IS will play while interacting with professionals.

7. *Integration of IS with existing software products*

Future software products will come with intelligence built into it. Vendors believe that the two industries will become integrated.

8 Potential Intelligent Machines

Some potential intelligent machines that have been identified in the interviews are described in other UNIS papers in these proceedings. For each machine, a brief description of its objective and capabilities are given, followed by a listing of such factors as the machine's requirements, elements of its configuration and other features [18], [32].

9 Components for Intelligent Machines

In this section, "components" are identified to realize some of the potential intelligent machines identified in the Phase I UNIS interviews. Components provide common platforms which can be used to drive the above machines. Although the machines serve different domains, intelligent systems can be designed around a relatively small number of components.

The technical feasibility of most of the components described will depend on the development of extremely fast processors, very high capacity direct access storage devices, and data channels with very high throughput. Overall, an estimate of improvements of orders of magnitude will be required for some to be cost-effective. Economic feasibility will require such improvements for acceptable cost/performance ratios. One specific need is for very fast document scanners with comparably fast optical character recognition software.

Innovative man-machine interfaces, both input and output, will be required. The basic purpose of many of the machines is to process inputs of very large amounts of text data. New ways of summarizing and conveying text data are needed. The Redundancy Suppressor described below is one example.

One sorely needed technology is speech recognition, capable of capturing continuous speech without training, and translating it into computer sensible form.

Many of the machines will require close man-machine interaction to solve a particular problem, perhaps requiring multiple "case studies." The concept of "case study" of verbal information must be explored, and Case Management Systems developed.

The early experiences in using spreadsheet programs found most new users having difficulty with the combinatorics of testing the interactions of multiple parameters. This soon became overwhelming. A CMS is a systematic way of dealing with multiple cases: storing them, comparing them, tracing the history of the development of the final case, etc.

A recent example of CMS for numerical cases is the MathProTM developed by MathPro Inc. of Washington D.C.

Potential components identified to date include:

1. SEMANTIC/SYNTACTIC ANALYZER (SSA). The SSA extracts meaning from text by analyzing both syntax and semantics. Its output is a set of assertions in some small number of standard formats suitable for further analysis and manipulation. For example: The probability that fact A is true is P. Event B happened between time t and time u. Data element C, required by process D, is missing.

2. RULE MAKER (RM). The purpose of RM is to create the rule base of an expert system without asking an expert explicit rule-oriented questions. It uses the output of SSA (based on whatever documents and other data are available) to create rules which can be used by an inference engine.

3. CONSISTENCY CHECKER (CC). CC analyzes the output of SSA for logical inconsistencies. It resolves them internally in so far as it can and reports what it has done. It refers unresolvable contradictions to human operators.

4. TREND ANALYZER (TA). The purpose of the Trend Analyzer is to predict the evolution of events which are described verbally rather than numerically. It is analogous to statistical techniques such as Box-Jenkins, multiple regression, etc. It uses the output of SSA to predict the likelihood of future events. It analyzes trends and pinpoints outliers; it describes the consequences of both likely and unlikely outcomes.

5. OPTIONS GENERATOR (OG). OG uses the results of SSA (as modified by CC) to create declarative statements which can be understood by humans, and can be used to drive word processors and other programs.

6. SYNERGISTIC INFORMATION COLLECTOR (SIC). SIC examines user requests for information and extends the requests to related subjects by identifying topics and concepts retrieved by the original request but which were not an explicit part of it. The process can be continued recursively to find "topics related to related topics" as far as required.

7. REDUNDANCY SUPPRESSOR (RS). RS examines sets of documents, identifies redundant information, and reports nonredundant information together with memoranda describing suppressed materials.

To date, seven components have been identified that could drive a set of intelligent machines. For example, intelligent system number eight ("Hopes and dreams extraction machine") uses SSA to analyze internal position papers, client statements, and public documents. It uses CC to check for inconsistencies and to request clarifications. It uses OG to generate statements of results, and it uses SIC to identify topics or concepts which are not an explicit part of the original request. This intelligent system might require a high level of customization, human interaction and exotic software.

Many machines described above have the following features: An Input/Output interface module which interacts with the users, accepting data from different media such as text, voice and video and providing information back to the user in different forms such as graphs, charts, or voice. The input/output interface module interacts with the knowledge base module. The knowledge base module serves as a repository of information. The knowledge base interacts with the inference engine. The inference engine makes inferences from its knowledge base in response to queries from the user. The IS performs the following tasks:

1. Update the knowledge base. Data can exist in various forms and formats and need to be converted into information and stored in the knowledge base.
2. Solve a particular problem using information in its knowledge base.
3. Provide Alternate decisions in response to queries from users.
4. Provide explanation on the solutions that IS arrived at. (See Figure 1 in Appendix A.)

Based on user and vendor interviews, professionals in different domains modularize information into information available publicly, information which is local to their organization and information they have acquired personally. The Knowledge base module can be conceptually modularized into:

*   *"Global knowledge module (GKM)"*, stores information about a particular domain that is public knowledge. For example in a real estate law domain, federal real estate laws, information available from the courts about different cases and court decisions are all global knowledge.
*   *"Local Knowledge module (LKM)"*, that stores knowledge about a domain existing in a particular organization. For example a particular law firm's cases in real estate law.
*   *"Personal Knowledge module (PKM)"*, knowledge stored by an individual - for example, the real estate lawyer's own case files.

The three modules have distinct mechanisms for knowledge update. PKM is updated depending on the amount of information the user is exposed to in a particular domain - for instance, the number of real estate cases the lawyer deals with. The LKM is updated depending on the number of active users in an organization in a particular domain. An example would be the number of practicing real estate lawyers in the firm. The GKM update is influenced by external factors - for example changes in property law in a state.

The core aspect of many of the machines is the use of existing data in a particular form and format and conversion of it to information stored and to be accessed from the knowledge base. Components are the mechanisms involved in the conversion of data into information. For example, the Semantic/Syntactic Analyzer (SSA) uses text and background information in a particular domain and identifies assertions. This process of conversion of data to information is knowledge acquisition [13]. Current research in knowledge acquisition includes Case-based systems [38], Combining Rule-based systems and Connection systems [22], integrating different knowledge bases [31] and Case-based learning [39].

New products need to be developed for different parts of IS. User interface is a key part and needs to be as close to human cognitive processes as possible. Innovative user interfaces which use a combination of voice, data and graphics to receive and display information may expedite the acceptance of IS. Users have time constraints and prefer user interfaces which expedite the process of using the system for problem solving rather than having to learn how to use the system.

10 Market Segmentation for Intelligent Systems

The identification of common user segments is called Market Segmentation. Market Segmentation is defined as identification of a group of relatively similar users who have needs or responses that are different from other users [42]. Identifying a group of similar users can help: determine the size of the potential market, select channels of distribution that may be needed, estimate pricing structure, identify product positioning opportunities and guide market research.

A market can be segmented on any of the following bases [41]

1. Demographics or socioeconomic variables - type of industry, size of company, location of company, age and income group.

2. Operating variables - technology the company possesses.

3. Customer's technical capability.

4. Customer's purchasing approaches.

5. Situational factors - urgency, specific needs, size of order, usage rates.

6. Personal characteristics of decision makers.

IS use knowledge acquisition, representation, storage and retrieval mechanisms. IS can be referred to as "Knowledge Ware," products which market knowledge in a particular domain. Knowledge is interpreted in a particular manner by users depending on the background knowledge they possess in a particular domain. The expertise of the user in a particular domain provides a natural way to segment users. Other Segmentation variables can be used to further segment the market but expertise forms a primary basis of segmentation, because the knowledge users have in a particular domain characterizes the flavor of information they require.

The intelligence system developers have three marketing segments to address, based on the expertise of the user in a particular domain. The three segments are characterized in the following sections. The characterization is based on interviews with users, vendors and MIS professionals [27].

10.1 Novice Segment

1. *Profile*: Novice users are neophytes in a particular domain. They are interested in accessing, acquiring and verifying information in a domain. Novice segment users look upon potential IS as domain experts.

2. *Benefits Sought*:

    (a) Education and training in the domain.

    (b) Use of the IS as reference to expert opinion.

3. *Features Preferred*:

    (a) A schedule for learning. IS need to provide the user with progress reports and evaluation on the learning process.

(b)   A "playing field for ideas." Mechanisms which allow users to test users' ideas and provide results on the validity of the ideas in the domain.

(c)   User interface needs to be natural to the users' cognitive process and needs to provide "help" functions that can lead the users easily into the domain.

(d)   Thorough explanations on inferences made by IS.

(e)   Mechanisms that can change the profile of the user from one segment to the other.

4.   *Example*: IS domain: Real estate law.

Novice Segment: New recruits to a law firm wanting to initiate or change to a career in real estate law will be in this segment.

10.2 Expert Segment

1.   *Profile*: Expert users are domain experts. They are interested in updating and validating their domain information. Expert segment users look upon IS as expert "associates."

2.   *Benefits Sought*: Update their knowledge in the domain. Follow changes in the global knowledge module (GKM) of the IS knowledge base. Validate concepts and ideas in a particular domain.

3.   *Features preferred*:

(a)   A "playing field for ideas." Mechanisms which allow users to test these ideas and provide results on the validity of the ideas in the domain.

(b)   Exchange of information between multiple knowledge base modules. Specifically, exchange of information between the Personal Knowledge Module (PKM) of the user and the three knowledge modules of the IS.

(c)   Explanation on inferences, on request.

(d)   Easy mechanisms to interface with IS which leads the user through the system.

(e)   Priorities to the use of different knowledge base modules.

(f)   Easy updating of knowledge bases of both personal and intelligent systems.

4.   *Example:* IS domain: patent law.

Expert Segment: Patent lawyers with many years of experience.

10.3 Associate Segment

1. *Profile*: Associate users are neither neophytes nor experts in a particular domain. They are users who possess a fair understanding of the domain but do not have extensive domain experience. They are interested in accessing, acquiring and updating information in a particular domain. Associate segment users look upon IS as expert colleagues.

2. *Benefits Sought*: Update their knowledge in the domain. Validate concepts and ideas in a particular domain. Enhance personal knowledge module (PKM) of user's knowledge base.

3. *Features preferred*:

    (a)   A "playing field for ideas." Mechanisms which allow users to test their ideas and provide results on the validity of the ideas in the domain.

    (b)   Explanation of inferences on request.

    (c)   Easy mechanisms to interface with IS which leads the user through the system.

4. *Example*: IS domain: Real estate law.

    Associate segment: Lawyers practicing environmental law but need information on real estate law.

(See Figure 2 for the table of characteristics of the three segments.)

Users' profiles change over time. A user initially in the novice segment may migrate to the associate or the expert segment. IS need to address profile changes in the design stage of product development. A different IS may be needed for each of the above segments or IS needs to be able to change profiles of users in real time.

11   A Methodology to Identify Potential Intelligent Systems

In the above sections, users' needs and vendors' perception of user needs have been analyzed, future intelligent components have been identified and a segmentation strategy to market intelligent systems has been discussed. From those discussions, the following methodology is proposed to identify and develop potential intelligent systems.

1. **Opportunity identification - Selection of DOMAIN:** Select a domain which the potential intelligent system will address. For instance, real estate law, health care planning or any other domain of vendor choice. The domain selection process is an

entrepreneurial process and depends on vendor's evaluation of their competitive advantage in the particular domain.

2. **Consumer measurement - Determine user needs:** After identifying the domain, interviews should be conducted with users in the domain about "user needs." The cognitive probing process described above can be used to determine problem areas potential intelligent systems can address.

3. **Identify features:**

   Each of the potential intelligent systems identified should be decomposed into their components and functional features. Functional features and component identification can provide estimates of the development effort needed to realize the intelligent system. The process enables vendors to identify modules of IS which can be reused from previous development efforts.

4. **Determine knowledge levels to provide:**

   The system should address the three modules of knowledge - namely, GLOBAL knowledge (public information available on the domain), LOCAL knowledge (knowledge different business entities possess) and PERSONAL knowledge (knowledge individuals possess) - and determine the parts of knowledge in the domain the intelligent system will support and provide. Knowledge elicitation can be a time-consuming activity. Therefore, time estimates for the development of each of the knowledge modules need to be undertaken.

5. **Models of consumers - Identify user segments:**

   Interview potential users in the domain and identify the different segments - namely NOVICES, ASSOCIATES and EXPERTS. Their individual needs for information can correlate with the knowledge levels identified in the previous step. For example, NOVICE users may be interested in PUBLIC knowledge while organizations they belong to may be interested in a combination of the three knowledge levels. EXPERTS may be interested in a combination of knowledge levels or mechanisms to access their PERSONAL knowledge. Interviews with users may include questions which can help demarcate the different segments in terms of profitability, needs, benefits to be provided and users' perception of the product.

6. **Predict market behavior:** From user interviews, estimate the number of potential users in each segment, the price users are willing to pay for the IS, and comparison of this IS with others having similar features.

7. **Evaluate new product development:** The decision to develop an IS or not is made on the basis of cost of development, technology constraints, market prediction and organization constraints.

8. **Refinement:** After the decision has been made to develop an IS in a particular domain, the IS is refined, based on the input from the users and the type of knowledge modules required.

9. **Fulfillment of user needs:** After refining the IS, the actual development of the IS is undertaken. Systems development and marketing work together to produce the IS. Users are presented with the actual product prototype and see how the IS performs relative to its promises. Based on users' feedback, the decision to market the IS on a large scale is decided.

(See Figure 3.) The output of the above iterative process helps identify

- Potential markets and target group of users.
- Benefits the IS offers to the users.
- Position of the IS versus its competition.
- Physical characteristics of the IS to fulfill user needs.
- Price, advertising and distribution strategies.

12 Analysis of the Current Expert System Industry

12.1 Organizational Structure of Current Expert System Vendors

As mentioned earlier, Expert System applications require Expert System shell providers and Expert System application developers. A majority of the shell vendors interviewed have about forty percent of the staff working on research and development. Training and after sales support constitute about twenty-five percent and the rest are marketing and sales staff [27]. The shell vendors interviewed sell their products on different hardware platforms ranging from personal computers, mini computers like VAX, to main frames like IBM. The shell is a tool. The quality of the shell is determined by the ease of use and the flexibility it provides in updating knowledge [27].

A few of the shell vendors have formed consultancy groups consisting of personnel who are both experts in different domains, and also possess a through understanding of the shell and other Expert System development tools. These consultants form the second line of product developers. After the sale of the shell, these consultants, also known as "knowledge engineers," develop the applications which the customers need.

12.2 Marketing of Current Expert System Shells

Expert System shells are marketed in various ways. One way is direct marketing and advertising. The other is to display the products in AI seminars and symposiums. Most of the sales are by previous customer references. Expert System development requires a lot of software engineering [38].

The development platform for Expert Systems varies from $400 for simple P.C.-based shells to several thousands of dollars for a mini or main frame solution. The costs of knowledge capturing spiral, depending on the thoroughness of the product in terms of rules that accurately define the domain and the comprehensive nature of the Knowledge Base of the Expert System [27]. Expert Systems are usually customized to meet the needs of a particular customer [27].

After identifying a potential customer, the shell vendors provide a variety of help to the customer in developing the application. They provide training in the use of the shell and provide documentation on how to use the shell. They provide the company either with their own consultants or recommend consultants to help users in application development.

12.3 Identifying New Products for the Expert Systems Industry

1. **Opportunity Identification - Domain Selection:** Vendors identify domains and develop Expert Systems based on their expertise, profitability and users' need for a particular domain [37], [13], and [11]. Individuals or a group of people within an organization find the following reasons to develop an Expert System:

    • The need for precise information to make decisions in a particular domain.

- Bottlenecks in the flow of information.

They conduct a feasibility study for developing an Expert System within the user organization. The study may focus on

- Necessity for information in the particular domain to be made available to many personnel in the organization.
- Retirement of a senior member who possesses domain information.
- Recruitment of new personnel.
- Scarcity of domain experts within the organization.
- Reorganization.
- The necessity to make decisions based on complicated and heuristic procedures.

2. **Determine User Needs:** After the domain has been identified, by the vendor and/or user, a study is conducted to determine user needs. Interviews are conducted with potential users of the Expert System for their opinion on the functionality the Expert System should possess. The cognitive probing process can be used to identify specific problems that need to be addressed by the Expert System.

3. **Identify Features:** The features for the 5 components of the Expert System are identified. For each component, the time needed to develop all the features are estimated. Tools needed to develop an Expert System are identified and evaluated. Organizations can use internal MIS personnel or can have external vendors develop the application.

4. **Knowledge Level to Provide:** A decision should be made on the amount of information the knowledge base is going to provide. The knowledge-base modules, namely: Global, Local and Personal should be identified and decisions should be made on how much information is going to be provided in each of the modules.

5. **Identify User Segments:** Interviews are conducted to recognize the different user segments, the features and knowledge level the segments need. Most domains can be segmented based on the expertise of the user in the domain.

6. **Evaluation of the product:** The number of persons in each segment, the features they need, and the cost to develop the different features are evaluated and a decision is made whether to develop or not to develop the Expert System.

7. **Refinement:** After the decision has been made to develop the Expert System, the features of each component of the Expert System are determined. Depending on the number of

user segments the Expert System is going to address, the knowledge level of the knowledge base is determined. After determining the knowledge level, the knowledge engineer has the task of getting information from domain experts and storing it in the different modules of the knowledge base. The time taken for development depends on the level of knowledge to be provided in the knowledge base and the availability of domain experts to provide the knowledge.

12.4 Other Issues in Current Day Expert Systems

Currently there are a lot of banks, insurance companies and Fortune 500 companies who have developed their own Expert Systems to help in their decision-making. The cost of development is high and companies have their own Expert Systems to solve particular problems. For example, two banks may have different Expert Systems in the domain of credit approval. Large corporations feel comfortable with applications tailored to their specific needs. This gives them control over the underlying rules that have gone into building the system. They are comfortable with the control over the knowledge base they have [27].

12.5 An Approach to Effective Expert System Development and Marketing

Expert Systems have to be domain specific in order to be useful and cost effective. Expert System shells are tools used to develop expert system products. In order to maximize profitability, many components in the development of Expert Systems should be reusable. In order for the components to be reusable, standards should evolve. User interface standards are evolving with windowing software. Most interfaces are click and control. Currently, there are no standards on how knowledge bases are organized, the mechanisms of inference engines, or means of providing explanations. Expert Systems have different standards for organizing their internal components [22].

A suggested approach to Expert System development and marketing follows.

- *Development Features*
    (a) Knowledge engineers or domain experts form an entrepreneurial venture to develop domain specific Expert Systems. The advantage of the venture is in the information provided to users.

(b)   The company buys components of the Expert System, such as the user interface development module, the knowledge elicitation and acquisition module, the explanation and results -- providing module from Expert System shell vendors.

(c)   The company tailors the expert system to meet the design criteria of the three user segments -- NOVICE, ASSOCIATE and EXPERT, while providing means to change profiles from one segment to the other.

(d)   The Expert System is marketed to users in the particular domain, with the user paying only for the access to the product.

(e)   The company undertakes to update the three modules of the knowledge base -- Global, Local and Personal.

(f)   Appropriate security mechanisms are developed to protect customer knowledge bases.

- *Criteria for Effectiveness*
  (a)   Access to diverse knowledge and different expert opinions for solving any problem. Providing the user with a better understanding of a problem.

  (b)   The cost of development of domain-specific solutions to be shared by multiple users.

  (c)   Customer saves money in terms of fixed costs, due to elimination of the need to own general hardware and software for Expert Systems.

  (d)   The knowledge updatating process is done by the Expert System company and the customer has access to up-to-date information.

  (e)   Due to cost advantages, the Expert System company can provide better customized features for different customers on high performance hardware and software.

- *Parameters (Constraints)*
  (a)   The Exert System needs mechanisms to infer from multiple knowledge modules. The means of providing access to multiple knowledge base modules is a technical problem currently.

  (b)   System life cycle will face progressive changes in the future. So the initial design should be flexible, to be modified at a later stage [15];

  (c)   Development of a flexible multiple segment Expert System is difficult; it's time consuming to develop a comprehensive system and constantly keep updating the knowledge base of the system [38].

(d)   Users do not want their proprietary knowledge to be used by others. Security mechanisms need to be developed to protect personal and local knowledge modules of the users in order for shared Expert Systems to be accepted.

(See Figure 6.) The approach describes a new industry called Knowledge Ware. Knowledge Ware developers are domain experts; their credibility depends on the thoroughness and ease-of-use of information they provide to users. The costs for individual users are shared, while providing secure mechanisms for information exchange.

13   Conclusions and Summary of Findings

Software systems, for example information systems, usually focus on analytical functions and provide faster means to improve efficiency and effectiveness of organizations. IS, in addition to the analytical functions, address users' cognitive needs to provide information which can aid the decision-making process of an individual or an organization. Expert Systems are Intelligent Systems in primitive form, with many missing features. Various components and engines need to be developed, in order to realize future Intelligent Systems which are most likely to be accepted by users. Expert Systems may provide the building blocks needed to realize future Intelligent Systems.

Users have difficulty conceptualizing new products for IS [27]. Therefore, methods like cognitive probing have to be used to identify user needs. Users are uncomfortable with systems which are black boxes, providing solutions without explanations [27]. Users prefer mechanisms which provide explanations on inferences [27]. Users demand mechanisms which can process data in variable form and format and provide directly usable information.

Future IS mandate knowledge bases which are substantially larger than those today. However, representing and acquiring knowledge is a difficult and time consuming task. Knowledge acquisition tools and current development methodology will not completely alleviate the problem of knowledge acquisition and make it go away, because the core problem is that knowledge is inherently complex and the task of capturing knowledge is complex. Thus knowledge currently acquired should not be wasted. Building qualitatively bigger IS will be possible only when knowledge can be easily shared by the development of standards for exchange of information between knowledge bases [31].

Most of the Expert Systems currently in use need personnel who translate the requests of the user into a specific Expert System format and translate system responses into a format easily understandable by the user. This procedure has a high potential for error in translation and leads to user dissatisfaction. Therefore, user interface will play an important part in the acceptance of IS. "Knowledge Ware products" need to provide interfaces which can lead the users easily and intuitively into the use of IS.

Many professionals like doctors and lawyers do not use computer systems for problem solving in their professional activity [34]. Professionals are skeptical of systems which address both the analytical and artistic part of the professional job. Any future systems which aid professionals in problem solving need to provide confidence-building mechanisms. These mechanisms help the user in better understanding the internals of the system.

Expertise of the user in a particular domain is a reasonable segmentation variable. Identifying the user's expertise level leads to better characterization and design of IS.

14 Future Research and Development

The cognitive-probing process identifies user needs by asking users questions about

* Their job function.
* How they solve problems.
* Training methods used in helping themselves or others to achieve increased professional competence.

The cognitive-probing process requires lots of time, effort and patience on the part of both the interviewee and the interviewer. The users may not be able to break their job functions into easy-to-distinguish parts or may have difficulty in identifying mechanisms used in training of others and themselves. The cognitive-probing process requires considerable effort on the part of the interviewee to identify user needs from the interview responses.

A possible solution would be to train the user in the cognitive-probing process and provide the user with a card in which they note the different activities they undertake during their work day. This card can be used to identify specific user needs. The problem in this method is that the user has to be constantly reminded to update the card. The card version of the cognitive probing process can lead to better understanding user needs [16].

Expertise is a natural segmentation variable. Once an IS is ready to market, other segmentation variables, such as purchasing behavior, will play an important role. The structure in the user organization will influence who would be the actual user of the IS and who will be the person deciding the purchase of the IS. The relationship between purchaser and user needs to be studied and this may lead to a different segmentation strategy.

A price discussion has not been undertaken, as IS do not currently exist in the form described earlier. Price is an important criteria. The pricing of shared access to IS can be an interesting research topic in the future.

IS requires large amounts of software engineering. The initial cost of development will be high [38]. Solutions for IS which can share the cost among many users may find increased support in the user community. IS may become increasingly more integrated with standard software. Future word processors, databases, and spreadsheets may come together with intelligent components [27].

The use of IS by professionals may be expedited or hampered by regulations. If IS in a particular domain become increasingly reliable then regulation may warrant the use of these systems by professionals in order to provide better service. On the other hand, if IS are not reliable, and professionals make decisions based on the results of the IS, this may lead to liability issues on who is responsible for the error, the professional or the IS.

15 References

[1] A. L. Lakshminarasimhan, D. Sinha: "Learning of Rules in an Expert System with a Probabilistic Expert", IEEE 8th Conference on Computers and Communication, Phoenix, March (1989).

[2] Buchanan, L., Feigenbaum, E.: "Dendral and Metadendral: Their Applications Dimension", *Artificial Intelligence* (1978).

[3] Bell, G.: "The Future of High Performance Computers in Science and Engineering", *Communications of the ACM*, 32(9), (September 1989) . Also: F. P. Agterberg: "Computer Programs for Mineral Exploration", *Science*, 245 (1989) 74-75. And: M. Yazdani (edt), *Artificial Intelligence Principles and Applications*, Chapman and Hall, New York (1986).

[4] B. G. Buchanan, E. H. Shortliffe: "Rule Based Expert Systems: The MYCIN Experiments of the Stanford Heuristic Programming Project", Addison-Wesley, Reading MA (1984) 59.

[5] Chicago Tribune: "Doctors Need to Use Computers, Panel Says", *Chicago Tribune Newspaper*, February 8 (1992).

[6] *Data Processing*: "Expert Systems Go to Work - A Survey of Tools for Getting Started in Expert Systems", April (1986).

[7] D. Davis: "Artificial Intelligence Goes to Work", *High Technology*, April (1987).

[8] J. Doyle: "Expert Systems and the 'Myth' of Symbolic Reasoning", *IEEE Transactions on Software Engineering*, 11(3), (1985) 1286-1390.

[9] R. O. Duda: "The Prospector System for Mineral Exploration", *Final Report, SRI Project 8172. SRI International*, Artificial Intelligence Center, Menlo Park CA, April (1980).

[10] Eric C. Ericson, Lisa Traeger Ericson, Daniel Minoli: "Expert Systems Applications in Integrated Network Management", The Artech House Telecommunication Library (1989).

[11] Eric C. Ericsion, Lisa Traeger Ericson, Daniel Minoli: "Introduction to Expert Systems", *Expert Systems Applications in Integrated Network Management*, 1989.

[12] Expert-System Resource Guide, *AI Expert*, May (1991).

[13] Morris W. Firebaugh: "AI A Knowledge Based Approach", PWS-KENT Publishing (1989).

[14] G. Loberg: "SMART II: Knowledge Requirements for Expert Systems", *IEEE ICC'88 Conference Record* (1988).

[15] G. Loberg: "Smart II: Principled Design of Knowledge-Based Systems", *Bellcore Artificial Intelligence Symposium Proceedings*, Bellcore, Livingston NJ, June (1988).

[16] Elie Geisler, "UNIS Phase II Status Report", *Center for Information in Telecommunication Technology*, May (1992).

[17] Gerry Hoffman: "Dreamware Machines" (1991).

[18] Gerry Hoffman, Nagaraja R. Srivatasan: "Components for Intelligent Systems", *Center for Information in Telecommunication Technology (CITT)* (1992).

[19] Linda A. Murray, John T. E. Richardson: "Intelligent Systems in a Human Context Development, Implementations and Applications", Oxford Science Publications (1989).

[20] M. Minsky: *The Society of the Mind*, Simon and Schuster, New York (1987).

[21] M. Minsky: "Communication with Alien Intelligence", *Byte*, April (1985).

[22] M. Minsky: "Logical Versus Analogical or Symbolic versus Connectionist or Near versus Scruffy", *AI Magazine*, summer (1991).

[23] Joel Moses: "Symbolic Integration", Ph.D. Thesis, Massachusetts Institute of Technology, Cambridge MA (1967).

[24] M. F. Deering: "Architectures for AI", *Byte*, April (1985).

[25] B. McNeil, S. Pauker, H. Sox, A. Tversky: "On the Elicitation of Preferences for Alternative Therapies", *New England Journal of Medicine*, 306 (1982) 1259-1262. Also: S. Weiss et al.: "A Model-Based Method for Computer-Aided Medical Decision Making", *Artificial Intelligence*, 11(4), (1978) 145-172.

[26] Nagaraja R. Srivatsan: "Interview Notes on Interviews with Professionals", May (1991).

[27] Nagaraja R. Srivatsan: "Notes on Interviews with Expert System Shell Vendors", May (1991).

[28] Allen Newell, John C. Shaw, Herbert A. Simon: "Preliminary Description of General Problem Solving Program-I (GPS-I)", *Report CIP Working Paper 7*, Carnegie Institute of Technology, Pittsburgh PA (1957).

[29] R. H. Michaelsen et al.: "The Technology of Expert Systems", *Byte*, April (1985).

[30] R. Reddy, L. Erman, R. Fennell, R. Neely: "The HEARSAY Speech Understanding System: An Example of the Recognition Process", *IEEE Transactions on Computers C-25*, (1976) 427-431.

[31] Robert Neches, Richard Fikes, Tim Finin, Thomas Gruber, Ramesh Patil, Ted Senator, William R. Swartout: "Enabling Technology for Knowledge Sharing", *AI Magazine*, Fall (1991).

[32] A. H. Rubenstien, E. Geisler, M. Glazer, G. Hoffman, J. Husain, O. Heller, D. Koester, N. R. Srivatsan, R. Walker, C. Wilkes: "Preliminary Design Features of a Series of Intelligent Support Machines for Professional Knowledge Workers", *PICMET Conference Proceedings* (1991).

[33] A. H. Rubenstein: "UNIS Idea Memo No. 4", *Center for Information in Telecommunication Technology (CITT)*, Northwestern University (1990).

FEATURES OF FUTURE INTELLIGENT SYSTEMS

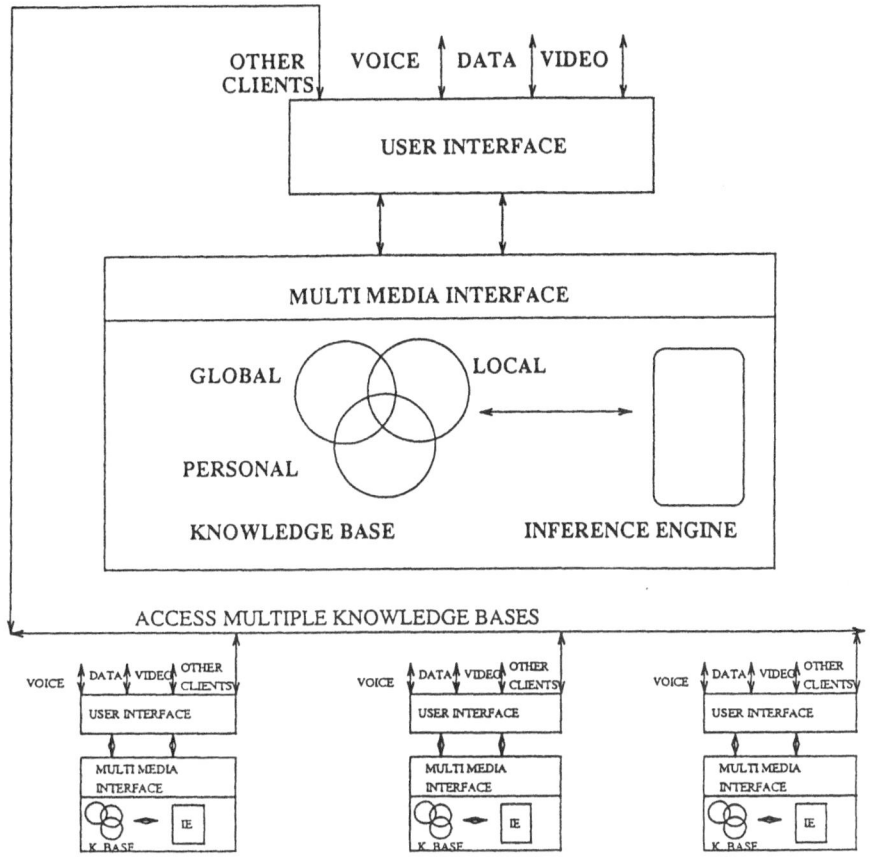

FIGURE 1

80

[34] A. H. Rubenstein, E. Geisler: "Users' Needs for Intelligent Systems (UNIS): First Year Progress Report", *Center for Information in Telecommunication Technology (CITT)*, Northwestern University, April (1991).

[35] R. Fennell, D. Richard, Victor R. Lesser; "Parallelism in Artificial Intelligence Problem Solving: A Case Study of Hearsay-II", *Tutorial on Parallel Processing*, IEEE Computer Society, New York NY (1981).

[36] R. Trappl: *Impacts of Artificial Intelligence*, North-Holland NY (1986).

[37] Ravi Sharma, David W. Conrath, David M. Dilts: "A Socio-Technical Model for Deploying Expert Systems - Part I: The General Theory", *IEEE Transactions on Engineering Management*, 38 (1991) 14-23.

[38] Roger Schank: "Where is the AI", *AI Magazine* 12(4), (1991), 39-47.

[39] Roger Schank, C. Ferguson, J. Barger, M. Greising: "ASK TOM: An Experimental Interface for Video Case Libraries", *Technical Report, 10*, The Institute for Learning Sciences, Northwestern University, 1991.

[40] K. F. Schaffner: "Ethical and Legal Issues Related to the Use of Computer Programs in Clinical Medicine", *Annals of Internal Medicine* 102 (1985) 529-536. Also: R. E. Dayhoff: "Medical Informatics: The Revolution in Law, Technology, and Medicine", *Journal of Legal Medicine* 7 (March 1986) 154. And: M. L. Robinson: "MD Malpractice Data Bank: Nobody's Happy", 62(17), (1988) 28-29.

[41] Thomas V. Bonoma, Benson P. Shapiro: "Segmenting the Industrial Market", D. C. Heath and Company (1983).

[42] Glen L. Urban, John R. Hauser: *Essentials of New Product Management*, Prentice-Hall, Inc., Engelwood Cliffs NJ (1987).

[43] Glen L. Urban, John R. Hauser: *Designing and Marketing of New Products*, Prentice-Hall Inc., Engelwood Cliffs NJ (1980).

SEGMENTATION GRID

FEATURES	NOVICE SEGMENT	EXPERT SEGMENT	ASSOCIATE SEGMENT
Benefits Sought	Education, Training, Learning mechanisms.	Knowledge update. Validate concepts & Ideas.	Knowledge update. Learning mechanisms. Understand domain terminology.
Features Prefered	Playing field for ideas. Explanation of inferences. Schedule for learning. Means to change profile. Extensive domain help.	Playing field for ideas Explanation on inferences Priorities in use of Knowledge Base. Intuitive user interface.	Playing field for ideas. Explanation of inferences. Intuitive user interface. Means to change profile. Schedule for learning.
Neccessary tools	Expert Guide. Learning tool. Reliable source of domain information.	Alternate decision paths. Global knowledge base. Second opinion system.	Learning tool. Training instrument. Reliable source of domain information.
Demography	Entry level person.	Domain expert with years of experience. Senior management.	Part of an organization in the particular domain. Middle management.

FIGURE 2

INTELLIGENT SYSTEMS: PRODUCT DESIGN PROCESS

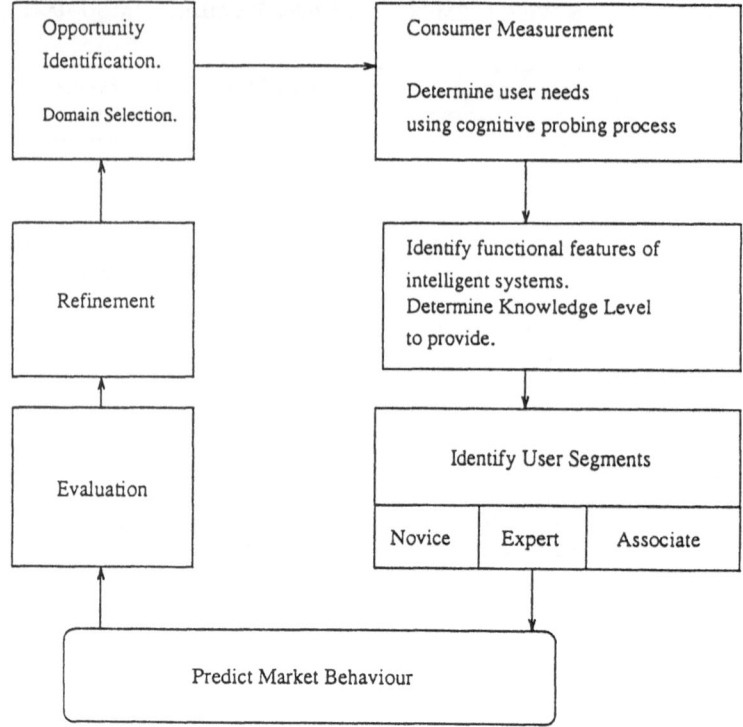

FIGURE 3

MAIN COMPONENTS OF EXPERT SYSTEM

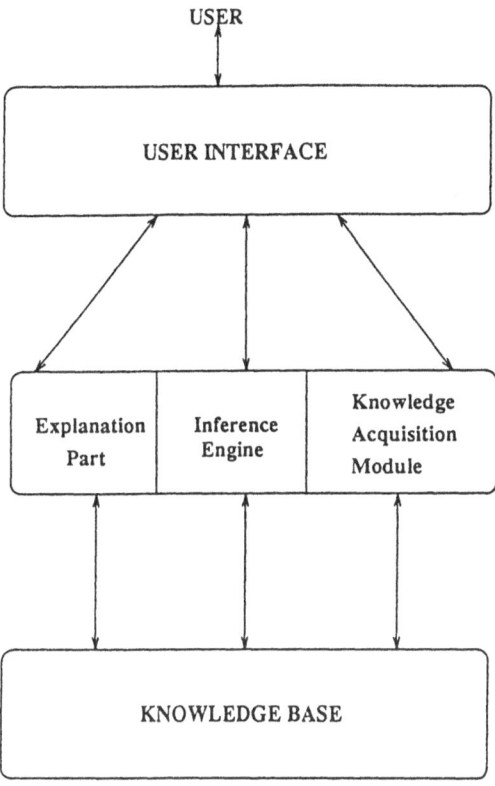

FIGURE 4

PROGRAMS AND PLAYERS IN EXPERT SYSTEM

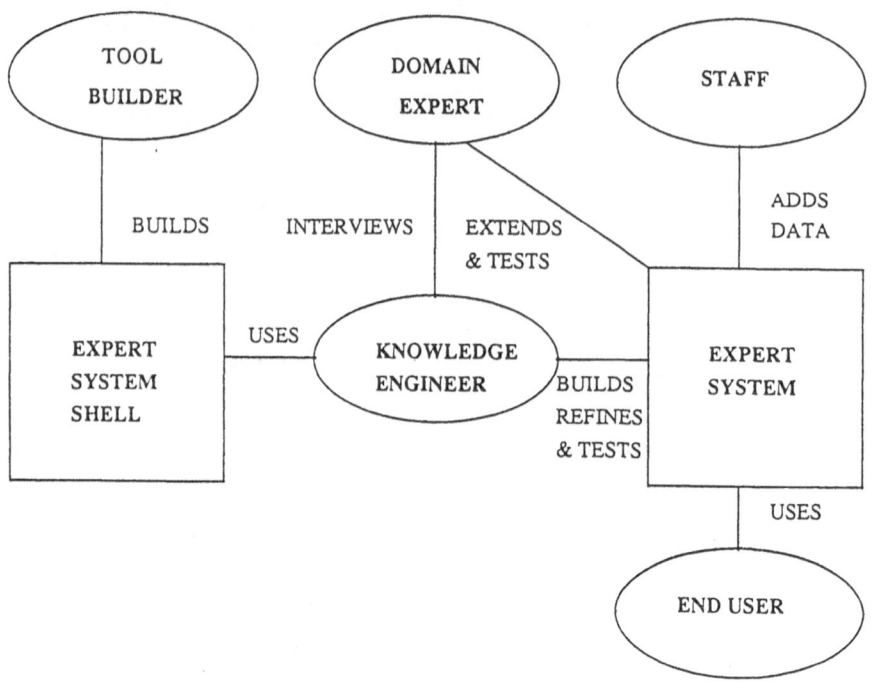

FIGURE 5

An Approach to Effective Expert System Development and Marketing

Design Features

Easy access to multiple knowledge bases

Smooth interaction between user and knowledge base provider.

Centralized organization of "Global" knowledge module.

Inference explanations.

Mechanisms to update personal knowledge base.

Efficient knowledge acquisition module.

Design Objective

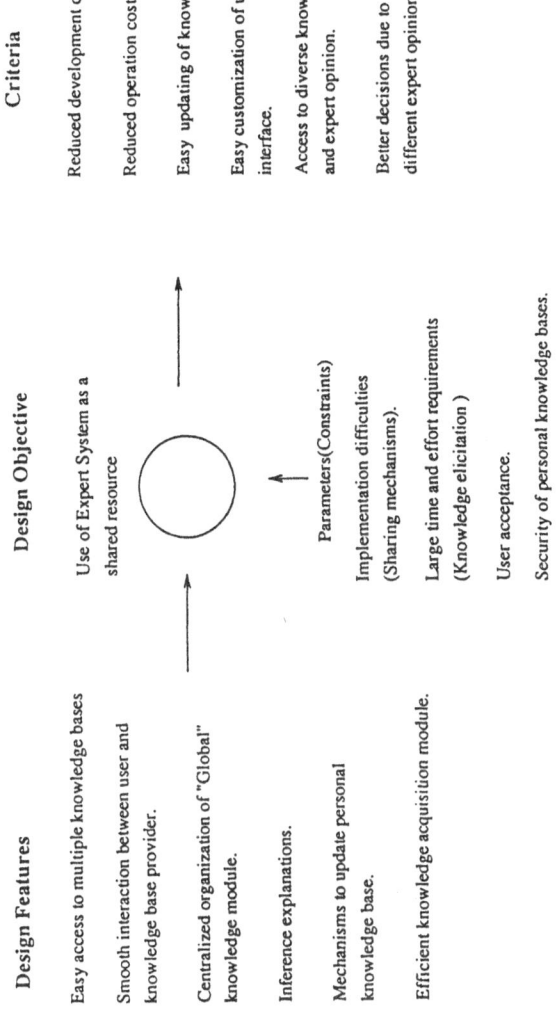

Use of Expert System as a shared resource

Parameters(Constraints)

Implementation difficulties (Sharing mechanisms).

Large time and effort requirements (Knowledge elicitation)

User acceptance.

Security of personal knowledge bases.

Criteria

Reduced development cost

Reduced operation cost

Easy updating of knowledge.

Easy customization of user interface.

Access to diverse knowledge and expert opinion.

Better decisions due to different expert opinions.

FIGURE 6

Telecooperation: Distributed Cooperative Work of Tomorrow

Jean E. Schweitzer
Siemens AG

Abstract

Cooperative working in a geographically distributed team is about to succeed. Local networked workstations are quite a normal situation in our daily working environment and even more and more the interconnection over public networks is established. From this it follows that the web of national and international business relationships is getting denser and denser. Worldwide contacts are getting closer. Important and quick decision-making require a powerful support to keep an intensive and close cooperation over long distances alive.

Practical and effective cooperation in today's worldwide expanded environment demands an efficient and integrated solution, i.e. a complete workstation with a single user-interface covering all kinds of media. Differently expressed, an arrangement of various tightly coupeled devices such as a PC with a separate video monitor, a telephone set, an infrared controlled recorder, etc. each having its own user interface requiring more or less the use of both hands are being rejected.

It is under this guideline we built the demonstrator MALIBU which is described in this paper. MALIBU shows particularly the hard- and software integration which creates a consistent and network capable multimedia workstation. Installing an infrastructure based on this workstation will constitute a multimedia telecooperation platform suitable for many domains of applications where the virtual elimination of distance may provide a considerable impact.

Introduction

During the last 5 years an important new interdisciplinary area has been set up under the name of "computer supported cooperative work" (CSCW). The main goal is to apply computer technology in order to make cooperation in workgroups more effective as they are at the moment. This of course is not all new, about 30 years ago several systems were concipated in order to help the people at their intellectual activities. The so called "cooperative Tools" were to extend the abilities of the people [Gre88],[Byt88].

Inspite of all advantages computer supported cooperative working has not yet been accepted on a broad basis, especially for a geographically distributed team there is no satisfactorily realisation. One reason is probably that certain technologies had not been sufficiently developed to make networked working easy enough for the user.

The current available technology for workstations, networks and user interfaces offers many opportunities for support of teamwork, but there is no generic framework and integration with existing applications is hard. If we link modern communication technologies (broadband, multimedia) with methods of distributed problem-solving of artificial intelligence (AI) and if we support this with models of cooperation which are analysed in the area of CSCW, then cooperative working receives a new dimension.

The Corporate Research and Development Department of Siemens initiated in front of this background a collaborative effort with the German Center of Artificial Intelligence. It aims at integrating CSCW approaches with modern telecommunication and network technology in order to use the synergetic potential of these two areas to create the basis for a new class of applications with high degrees of distribution and cooperation. This includes the modelling of telecooperation processes as well as their realization in a physically distributed multimedia infrastructure [Die90],[DFK92].

Cooperative Distant Learning/Training

An application scenario which is being examined is the area of training respectively interactive tutoring, especially internal further education and product training. It is supposed, that a course or a training unit will bring together participants from different locations, so that they normally have to overcome certain distances [Lux91],[Sch92].

The present scenario focusses on a new kind of "distant learning". In this scenario the participants will be supported in their desire of learning by a distributed learning environment composed of particular workstations attached to an appropriate wide area network. The application and system software implemented upon this physical infrastructure will provide a new form of interactive learning. This requires sufficient possibilities for interaction between the users and the teaching/learning material. Furthermore it is expected that the learning process will be conducted in a cooperative way applying all the necessary and available communication media. Such a scenario forms a sufficiently complex background to deduce system requirements which are also representative for other potential CSCW domains.

The research efforts are carried out under the project title MALIBU (Multimedia Active Learning In BUsiness-environments) One of the first milestones is the development of a demonstrator (see figure 1) which realizes basic functions of a distributed, cooperative system.

Due to the increasing complexity of the products and increasingly shorter product innovation cycles the requirements in the area of training have increased. Courses have to be made available in very short intervals that means the corresponding knowledge must be transmitted in a fast, direct, extensive, clear and efficient way. In most cases trainer and learner are in different locations therefore travel expenditure and time spent on these requirements increases. The problem is often avoided by using so called Computer Based Training (CBT) software.

From the learners point of view CBT-programs have the disadvantage of neither having contact with a trainer nor with other learners; the learner is all by himself. From the manufacturers point of view CBT-programs are rather expensive to produce. It is a basic rule that the quality of a CBT product must be the higher the more an autonomous learning is intended. Even with good preconditions e.g. a valuable authoring system it can easily lead to production costs of about 1 manmonth per hour CBT. For short-lived contents CBT-programs are scarcely offered due to economic reasons and therefore the CBT-software is rarely just in time.

MALIBU meets the mentioned grievances by applying technical innovations like broadband-networks multimedia-technology etc. In this sense MALIBU gives impulses for the construction of a learning environment of the future.

The variety in presentation forms of the training material based on the progress in the multimedia sector (audio/video conference, multimedia documents) may offer to the learner a high-grade type of distant learning. A given fact is to be adopted much better to the knowledge and to the receptivity

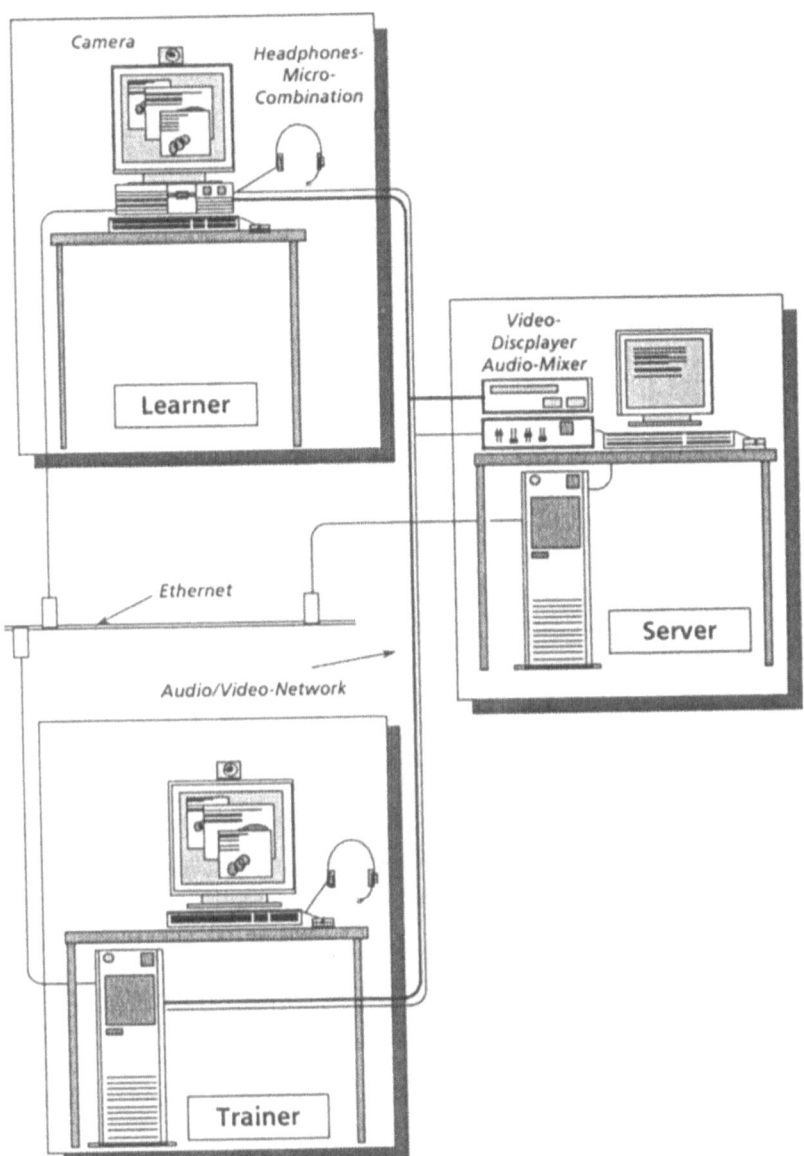

Fig. 1 MALIBU-Demonstrator

of the learner because of the skillfull order of text, graphic and fixed image elements with the support of audio/visual sequences.

The technical problems for the management of huge data amounts which come up when implementing multimedia technology can be solved. First efficient video compression algorithms for the reduction of data traffic have been developed with great research expenditure. With the regard to network technology existing local broadband networks as well as the public broadband networks which will soon be available deliver the possibilities to install full multimedia (text, graphic, still image and full motion picture and audio/video conferences) in a distributed broadband environment.

Compared with conventional computer supported training active distant learning in a distributed broadband environment means:

- faster spreading of (up to date) knowledge
- a flexible organization of the classes by decentralization; transfer of the classes into the regular working place of the learner, so that he can use his own office i.e. the problem of distance becomes a neglectable factor.
- better adoption of a given fact to the knowledge and receptivity of the learner by skilled arrangement of text, graphic and still image elements and with support from audio/visual sequences.
- individual training/supervision by the trainer, i.e. a face-to-face communication via audio/video conference with simultaneous cooperative processing of multimedia documents.
- higher efficiency due to cooperation with other learners also under the mentioned conditions.

With active distant learning two kinds of cooperation can be distinguished:
1. Cooperation between trainer and learner
2. Cooperation within a group of learners

A cooperation between the trainer and each learner is always bilateral. On the one hand after the presentation the learners can always ask questions concerning the content of the lesson or to clarify points that were not quite understood. On the other hand the trainer can either during or after the presentation of the lesson ask questions or give exercises in order to test or see the reaction of the learners.

Apart from this there is the opportunity for cooperation between several learners so that an exercise can be dealt with in teamwork. In doing so the learners can help each other and they can solve more complex problems. In this case the trainer has mainly an advisory function and can be considered as a member of team.

Besides the so called human agents (trainer, learner) machine agents also play a role in this new form of distant learning. The machine agents, for example a technical documentation system, an optical disk drive connected to a server or a conference coordinator support the human agents at the dealing with their lesson.

The MALIBU-Demonstrator

On the basis of networked UNIX- Workstations, Siemens-Nixdorf-WX 200, MALIBU shows the following highlights for the support of cooperative working:

- An audio/video conference integrated into the workstation for the communication between the team partners.
- Including of video sequences via remote control from a distant server.
- Joint Editing for cooperative editing of a document i.e. several team partners can work at the same document from different workstations.

The spectrum of implementation of such networked configuration reaches from simple teamwork on a joint document over the here chosen application scenario of the field of training to the complex tasks of cooperative project management. The chosen scenario forms a sufficiently complex background in order to deduce system requirements which are also representative for other cooperative domains.

The system presentation to the user (e.g. the user interface) is important for accepting telecooperation as a new kind of work practical and effective cooperation in today's worldwide expanded environment demands an efficient and integrated solution, i.e. a complete workstation with a single user-interface covering all kinds of media. Differently expressed, an arrangement of various tightly coupeled devices such as a PC with a separate video monitor, a telephone set, an infrared controlled recorder, etc. each having its own user interface requiring more or less the use of both hands are being rejected. In MALIBU especially the hard/software integration is considered so that the result is a compact multimedia workstation.

To be able to integrate all the functionallities into one user interface, we chose the X11 window-system and the look-and-feel of OSF-Motif, provided by the MOTIF-window manager and the Motif-Widget-sets and -libraries for the application programer.

The integration of full motion video in PAL-resolution is performed by a combination of high-resolution graphic- and frame-grabber-boards, controlled by a X11-server with a special video extension. This video-extension, derived from results from the ATHENA-project at MIT, and the features of the frame-grabber board enables video-windows, which can be moved, resized, iconified etc. as other X11 too. The video signals from the two inputs, which are switched by software, can be displayed alternately as live or still video, e.g. in one window runs a training video, in the other you have the frozen picture of your trainer. To get a snapshot of the video is easy to do - click the freeze-button in the control panel of the video-window and save it later as a window-dump-file. To get it back on screen is analogue - undump the file and see, what it contained.

Joint working with the MALIBU demonstrator is based on the shared-X system. The SharedX system was developed originally by DEC (CEC Karlsruhe) from M. Altenhofen and others. The first public version of SharedX was distributed as public domain software and stored on many FTP-servers.

SharedX ´s principle functionality is shown in figure 2. The X application sends its graphical output to the pseudo-server part of the SharedX bridge. The output is then multiplied and sent to the pseudo-client parts inside the SharedX bridge. These pseudo-clients theirselves transform it once more according to the requirements of the X servers, they are connected with, and at last sent the output to them. In the other direction, the users´inputs are transmitted from the X servers to the appropriate pseudo-clients in the bridge. There, all the input coming from the user who has the "token" is transformed and sent further via the pseudo-server to the application. The other users´ input is dropped at the related pseudo-clients.

Some imperfections have been found in the time of practical usage of SharedX for our purposes. Some of them have been fixed.

The single user application FRAMEMAKER, a desktop publishing system running under X11, is shared via the shared-X bridge to at least two partners, sitting on different machines, forming a so called joint working session. FRAMEMAKER´s output is multiplexed to all the partners in the session. The input signals, coming from the mice and the keyboards of the partners in the session are scheduled in the shared-X bridge according to the token-holder-entry in the X-properties. These properties, which are held consistant at all locations by the X11 system, can be modified by local session-management

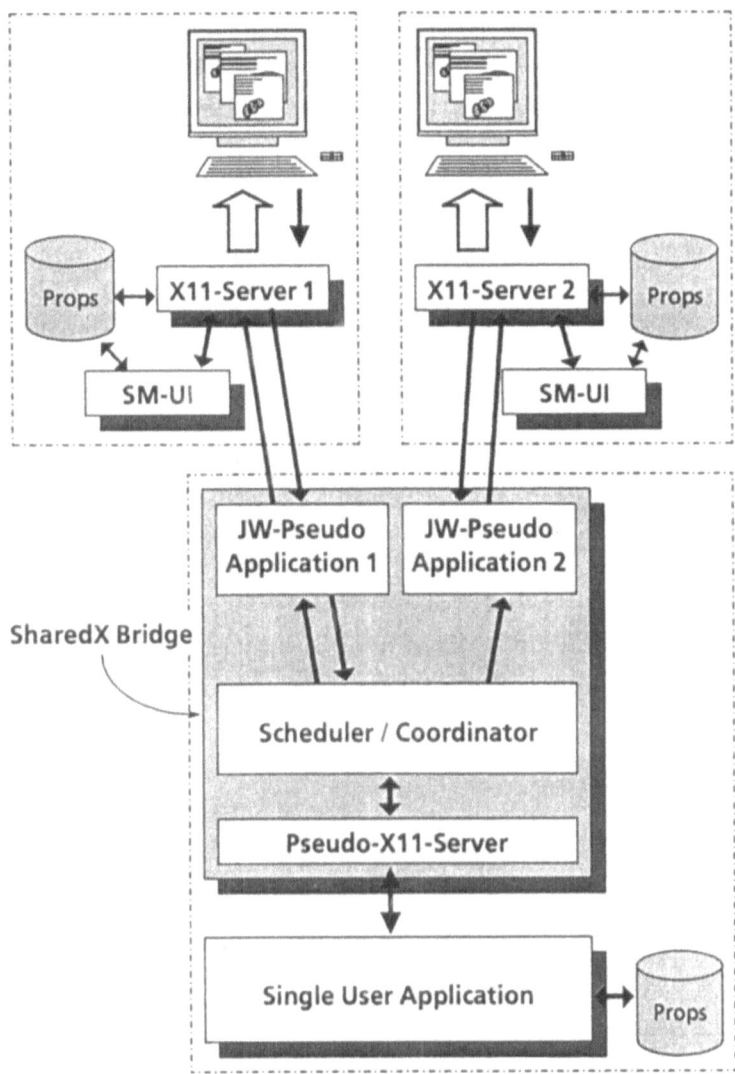

Fig. 2 The shared-X principle

interfaces (SM-UI), so that passing the token, e.g. the permission for editing the shared document, is only clicking on the partner, listed in the SM-UI. Joining and leaving the session is as easy as this - ask the session leader by an audio/video conference call or send him a simple message, and he will add you to the session-partner-list.

The HW/SW-configuration, used in the MALIBU-demonstrator is shown in figure 3.

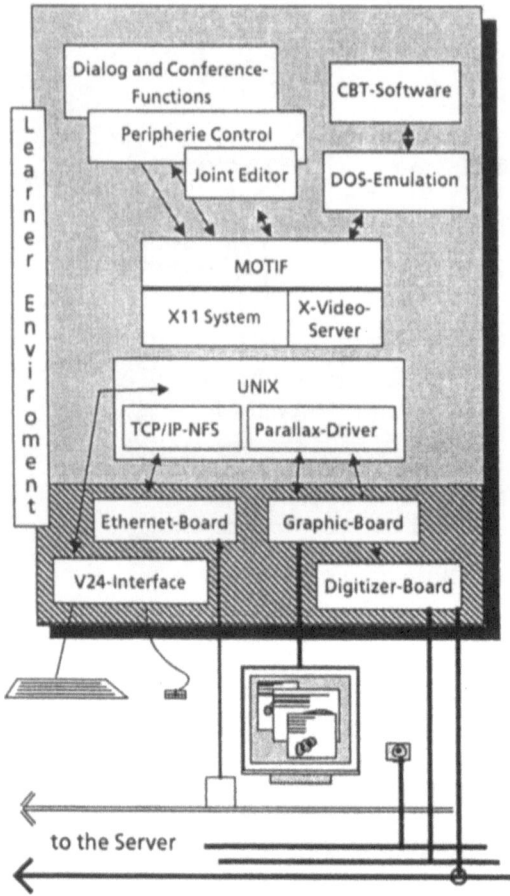

Fig . 3 HW/SW-Integration (Learner-Workstation)

Further investigations

Real teamworking and the related requirements to the supporting infrastructure become obviously clear for a team larger than three partners. So the decision was to extend the MALIBU demonstrator to a multiperson environment.

In a first step which is achieved now we expanded the local hybrid network (a combination of Ethernet-LAN and analogue lines served by an audio/video switch) to a capacity of 8 multimedia workstations. Every participant can now communicate and cooperate with several partners; the interaction relationship grows up from 1:1 to n:m.

The increased effort in cabling and the separated switching of the different information classes is more then compensated with the advantage of the immediate availabilitity of the multimedia communication facilities. This is of great importance for the progress of our investigations because it provides an early experimentation platform on top of which we could install and evaluate new telecooperation services .

We enriched our on FrameMaker based telecooperation platform with several applications and tools like a cooperative distributed game, a cooperative ressource sharing manager and an audio/video conference tool with an interactive grafical user interface. Besides this own developments we applied some other commercial x_applications on the window sharing mechanism. From this it became clear that the actual functionality is not sufficient and we designed therefore a more complete architecture [Cron92].

The European Community is founding under the frame of RACE and DELTA several project where the focus lies on distributed cooperative work. We are involved in corresponding tasks within the projects ARAMIS, CIO, ECOLE and MALIBU. According to the requirements of the application domains the underlying digital network is

- the narrowband ISDN; in this case the availability has a higher priority than the bandwith, so that a reduced multimedia functionality can be accepted.
- or the ATM broadband network; here full multimedia functionality is required and the project has a prototypical meaning

How to integrate telecooperation services into a user interface is for further investigations. Until now it is not yet clear that kown interface metaphors can satisfy the new requirements which distributed cooperative

One of major advantages of the applied window sharing mechanism is that any particular x_application can be shared without modifying the application itself (application transparency). A telecooperation platform based on this mechanism is called a CSCW transparent system. Beside this we have to distinguish the socalled CSCW aware systems. Here a modification or even a new coding of the application is necessary and this requires a considerable programming effort. On the other hand their is no functional restriction. The degree of flexibility is high and the multiuser functionality can largely be adopted to the applications need [Jar92]. We suppose that an optimal system will have a mixture of CSCW transparent and CSCW aware components and therefore we follow both directions.

Tools for constructing a telecooperation environment a very rare, but their will be soon a considerable demand. For more detail concerning our work on this topic we refer to our second paper in this volume [S-H92].

Acknowledgment

The author thanks the TEAMKOM-Group for valuable commentary in writing the paper and for active help in the realization of the demonstrator, namely C. Dietel, B. Otto, W. Reinhard. D. Scheidhauer A. Scheller-Houy, and T. Schmidt. The author also thanks the students of the university of Saarbrücken for their active work in the MALIBU-Project

References

[Gre88]   I. Greif (ed.); Computer-Supported Cooperative Work: A Book of Readings;Morgan Kaufman Publishers; 1988.

[Byt88]   BYTE; Groupware; P. 242-282; McGraw-Hill Publication; December 1988.

[Die90]   C. Dietel et al.; KIK-Projektbeschreibung, Version 2.0; September 1990.

[Lux91]   A. Lux, J. Schweitzer; MALIBU: Interaktives kooperatives Arbeiten in verteilter Multimedia-Umgebung; Berichte des German Chapter of the ACM; Band 34; Teubner Stuttgart 1991

[DFK92]   DFKI; Wissenschaftlich-Technischer Jahresbericht 1991; Document D-92-15; April 1992.

[Sch92]   J. Schweitzer; MALIBU: Interactive Cooperative Work in a Distibuted Multimedia Environment; Proceedings of the International Workshop on Advanced Communications and High Speed Networks IWACA 92, Munich 1992

[S-H92]   A. Scheller-Houy, R. Bartels, D. Scheidhauer; Knowledge based Telecooperation; in this document

[Cro92]   S. Cronjäger, W. Reinhard, J. Schweitzer; Functional Components for Multimedia Services; to be published at ICC

[Jar92]   A. Jarczyk, P. Löffler, G.Völksen; Computer Supported Cooperative Work (CSCW) State-of-the-Art and Suggestions for Future Work; Siemens Internal Report; Munich September 1992

Assisting Systems for Processing Inexact Knowledge
The TASSO Project: Overview and First Results

Franco di Primio

GMD, FIT.KI

email: diprimio@gmdzi.gmd.de

Abstract

The GMD key project "Assisting Computer" (AC) is looking for a new 'division of labor' between human and computer. Previous research done in the context of the knowledge-based paradigm aimed mainly at increasing the domain competence of software systems. The AC additionally addresses the goals of supporting cooperative (not only individual) work and of providing "modally" adequate assistance, which means that the focus lies on *how* the system is expected to use and apply its domain knowledge. It should, for instance, be able to explain its behavior when desired, it should adapt itself to the needs and personal style of the user and it should be able to cope with imprecise instructions on the basis of background knowledge or of current and previous contexts. In this paper we will concentrate on this last point and outline some of the results we are going to achieve in the TASSO[1] project (TASSO is an inexact acronym for Technical Assisting Systems for Processing Inexact Knowledge). Our basic research goal is to study and develop new methods for handling incomplete, vague or contradictory knowledge. From a communicative or interactive point of view, which is relevant for the quality of the assisting modalities, it is important that application systems incorporating these methods be able to interpret inexact user instructions. As an example application field, we are currently studying the problems of processing imprecision during the construction and search of graphics. The user can incrementally specify the attributes of graphic objects (such as presentation graphics or diagrams of office furniture layouts). Adequate assistance consists here in completing and making more precise or consistent user specifications that are inexact in the above sense. The techniques we are studying and applying are mainly based on non-classical inference, such as non-monotonic and associative reasoning, and on methods of planning under uncertainty.

---

[1] The project TASSO is partially funded by the German Federal Ministry for Research and Technology under number ITW 8900 A7.

Preliminary Remark

The GMD key project "Assisting Computer" (AC) is looking for a new 'division of labor' between human and computer. Rather than concentrating exclusively on the problem of how to increase and ameliorate the domain competence of software systems (this was the primary goal of research done in the knowledge-based context), AC focuses on the problem of how to improve the interaction between human and computer systems. The main intention is to complement and not to automate tasks completely. Technically this reflects the difference between open- and closed-loop systems [DG88, p. 4f]. Assisting systems belong to the class of open-loop systems: their constitutive characteristic is the fact that relevant control decisions remain and are taken outside by a human user interacting with the system. This is, of course, a very general property common to almost every interactive system. The distinguishing feature of AC is the goal of devising systems whose interactive behavior shows properties of an *effective* assistance. A human assistant, for instance, is not only expected to have sufficient competence in his domain of expertise, but also to be aware of his limitations, to be able to explain his own suggestions and to (re)act intelligently even on the basis of partially specified tasks. The technical realization of (aspects of) these properties clearly increases the quality of interactive software systems, which in their actual form are extremely limiting: the course of interaction is rigid and does not allow the user to leave or change the preprogrammed ways. While using a conventional system most resembles acting while dressing a straitjacket, the AC is expected to free the user and give him way to unfold. The total result should be a sort of creative symbiosis in which human intelligence and silicon computing power would each complement the other's capabilities.

1. The TASSO Project

In this paper we will concentrate on one assistance property: the ability to handle inexact knowledge and instructions[1]. Research on this point is done at GMD within the TASSO project (TASSO is an inexact acronym for Technical Assisting Systems for Processing Inexact Knowledge). The main goal is to study, contrast and develop new methods for handling incomplete, vague or contradictory knowledge. From an interactive point of view, which is relevant for the quality of the assisting modalities, it is important that application systems incorporating these methods be decidedly more flexible than conventional ones, being able, for instance, to interpret inexact user instructions. As an example application field, we are currently studying the problems of processing imprecision dur-

---

[1] For an overview of all other aspects of the AC project the reader is referred to [Ho91].

ing the construction and search of graphics. The user is expected to incrementally specify the attributes of graphic objects (such as presentation graphics or diagrams of office
furniture layouts). Adequate assistance consists here in completing and making more
precise or consistent user specifications that are inexact in the above sense. The techniques we are studying and applying are mainly based on non-classical inference, such as
non-monotonic and associative reasoning, and on methods of planning under uncertainty.

The study of non-monotonic inference mechanisms [Br91] is of central interest because
they allow the assisting system to handle incomplete problem specifications on the basis
of standard assumptions. Default rules expressing typical but not universally valid relationships help, for instance, to infer plausible conclusions and take meaningful decisions
in the absence of precise knowledge.

Associative reasoning techniques based on neural networks [He91] [Pa92] realize a strict
memory-based approach and are being used in this sense for the storage and retrieval of
known problem specifications and solutions. To solve a problem in this context means to
look at similar, already solved problems. These are re-activated in a content- and con-
text-driven manner.

Research on planning under uncertainty is motivated by its special role in the construction of graphics. The design of semi-standardized graphics like business presentations
can be viewed as a configuration task. In this context, existing techniques for planning
and configuration must be extended to allow the inexact specification of the planning
model, the planning operators and the goals [Her91].

Finally, since a problem specification may be inexact in every respect, i. e. incomplete
and/or vague and/or contradictory, the problem arises of how to apply the different
methods and inference techniques in a coordinated and complementary manner. This we
expect to be a problem of designing an effective and efficient hybrid architecture, i. e.
one allowing to use, at the right moment, the right method for the right (sub)problem.

In the following sections we outline the main results we have so far achieved in the study
of non-monotonic and associative reasoning techniques. In this presentation, we are
going to take a purely functional perspective, meaning that we want to stress the aspects
of the methods that are relevant from the point of view of an end user omitting the
(considerably more but partly tedious) details concerning the realization of the techniques (for which the reader is referred to the technical project reports). Furthermore, we
completely skip the research on planning under uncertainty and the architectural problem because, at the moment, we have not yet achieved any "practical" result in this
respect.

1.1 Why non-classical inference methods

The basic motivation for studying and applying non-classical inference methods lies in the strong need for flexibility in order to handle inexact knowledge. This is best understood by reference to the notion of a frame, which, as a major type of knowledge representation structure, has been, toward this goal, extremely influential in artificial intelligence [Mi80]. The premise underlying the frame idea is that knowledge about regularities in the world (such as the likely properties of objects and situations) is stored in the memory in clusters that can be accessed by need as large, coherent units and that can serve to generate plausible inferences ("expectations") helping to fill missing details in a given situation. Frames are paradigmatic for the idea of memory-based reasoning, i. e. reasoning which is not the result of sophisticated logical inferences but rather of recalling and exploiting known situations. Perhaps, their value is best summarized by a (provocative and paradoxical) sentence of C. F. von Weizsäcker, "Man sieht nur, was was man weiß" (We see only what we know)[1], that, in our context, can be used to emphasize the intended expectation-driven function of frames in cognitive processing. This function is crucial as it allows to complete by 'default' values (that are assumed unless explicit information is supplied to the contrary) partially specified or accessible situations. As devices for chunking information, frames have, indeed, originally inspired a bunch of productive research and application in default reasoning. Although they have demonstrated to be very valuable in formalizing knowledge about recurrent and stereotypical situations that can be used for 'completion' purposes, they show, however, some deficiencies in other respects. They are, for instance, not very flexible and sound in handling exceptions. What is to be done, in the course of a problem solving process, with objects or situations which are not properly matched by any known (stored) frame? Another problem can be called the "deactivation" (or "revision") problem. Suppose a frame has been activated (accessed) in response to partial aspects, i. e. some features of a given situation, and suppose that it turns out to be the "wrong" frame (i. e., the expectation is not met). How can it be deactivated? This is a complicated problem when you consider that frame activation technically can 'trigger' the execution of several (so-called attached) procedures that may cause, as a side effect, the attributes of the actual and other frames to be changed in form of a "chain reaction".

---

[1] We quote from [Bi89, p. 79]. It is worth noting in passing that the german word *wissen* (to know) is related to the latin verb *videre* (to see). The sentence then reads: We see only what we saw, i. e., in this respect, to see can be considered to mean to remember (think of Plato's teaching) what one has already seen. Currently, the notion of 'perception' is still problematic. Wittgenstein's distinction between *seeing* something and seeing it *as* something as well as the *theory-ladenness* objection of Thomas Kuhn (s. [BA91, p. 159f]) strengthen, for instance, the analysis that what we see depends upon what we know, and considerably challenge the notion of objective observation jeopardizing, in this respect, the epistemological status of science.

1.2 The EXCEPT system

Logically, the activation (or instantiation) of a frame corresponds to making assumptions or asserting some sentences. In this view, the deactivation problem just mentioned consists in undoing the effects of activation, i. e. retracting the sentences and all their consequences. The reasoning system EXCEPT [Ju91] has beeen conceived and developed with the explicit goal of correctly handling the completion problem as well as the exception and retraction (deactivation) problem in a logic-oriented framework.

The system provides a basic representation formalism which is almost as powerful as a first-order language. Compared to the well-known Prolog programming language [CM84], not only definite but full Horn clauses (true negation) are supported. It is even possible to define general clauses (containing more than one positive literal) with the restriction, however, that variables appearing in positive literals of a clause must also be used in at least some negative literal of the same clause. In addition to this language, which makes it possible to write formulas having the character of premises, the user is also allowed to qualify formulas, i. e. to characterize them as having the status of *assumptions* or *defaults*. The difference between assumptions and defaults is as follows. Assumptions are possible hypotheses that are introduced in order to explain observations. Given , for instance, the premises $a \rightarrow b$ and (the observation) b you can hypothesize (abduce) a as an explanation of b. Abduction is clearly unsound from the deductive point of view but a useful form of common sense reasoning. For an accurate analysis of the relationship between abduction and deduction the reader is referred to [Con91]. Defaults are conjectural rules (of thumb) which are applied whenever there is no evidence against their application. An (abstract) example: *Generally, objects of type T have property P. If A is an object of type T, then, in lack of any counterevidence, you can 'deduce' (expect) that A has property P.* Compared to abductive assumptions, the basic function of default rules, which has probably also to do with some form of 'cognitive economy', is different. While people, normally, trie to minimize the number of hypotheses (assumptions) needed for explaining something, they trie, in their actions or 'expectations' about the surrounding situations, to maximize the number of applied defaults (this reduces the amount of missing information). Assumptions as well as defaults (like 'prejudices' in everyday life) are used frequently in the reasoning process but are (hopefully) withdrawn if further premises become known, that contradicts some of them or some of the consequences based upon them. The internal machinery used to this end by the EXCEPT system is a rule interpreter extended by reason maintenance capabilities. This combination (reason maintenance plus a system for applying rules with additional tests) is a distinguishing feature of EXCEPT and has proved powerful enough to run on the system not only various forms of default logic but also other types of non-monotonic logics like, for instance, autoepistemic logic [Ju91, p. 37f][1].

From the user's point of view, the interesting aspect is that assumptions and defaults, of course, are not unrelated. They may depend on each other. They can, for instance, be more specific than others or conflict with others, where exceptions are a particular form of conflict. To understand the general perfomance of the system, it is useful to consider the set of premises together with the set of formulas expressing assumptions and defaults. To the same degree as assumptions and defaults are in conflict with each other and the premises, they make the resulting set (union) inconsistent. From this abstract point of view, the output of the system consists in partitioning this set into subsets which are maximal with respect to consistency and which satisfy the additional restriction to contain as a subset the set of premises, i. e. as much of the available exact information as possible. Each of these maximal (largest) consistent subsets (MCS for short) represents then a possible completion of the premises. This can be graphically summarized as follows:

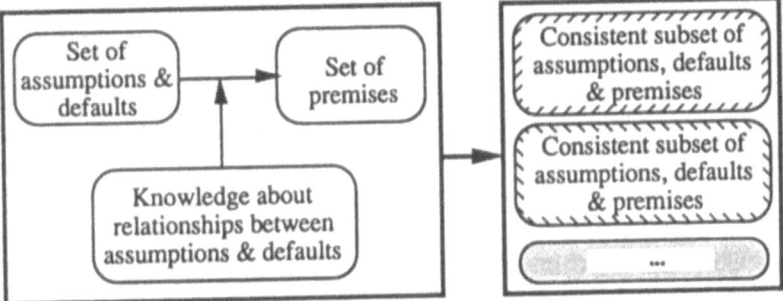

Controlling the extension of knowledge

Here the knowledge about the relationships between assumptions and defaults (including priorities or partial orders (preferences)) is used to control the extension of the set of premises through instances of defaults and assumptions. This (second-order) knowledge can be expressed as a set of meta-predicates (a meta-language) ranging over formulas of the basic representation language (the object language). The performance of the EXCEPT system can then be viewed and explained logically in terms of the properties of the resulting amalgamated language.

From an incremental and interactive point of view one can see the process of constructing MCS as maintaining multiple contexts, where each context represents an alternative

---

[1] Autoepistemic reasoning is a further form of non-monotonic reasoning that is revisable because it is of introspective nature (*From my current state of knowledge, I can deduce that ...*), i. e. it is 'valid' only with respect to the state of knowledge of a particular reasoner, which is subject to change [Tha88, p.165ff].

completion of a partial problem specification to be offered to the user as a choice. A satisfactory management of multiple contexts, including the possibility for the user to undo the effects of previous choices, is not an easy task. It is, however, essential because, for instance, it allows the user to explore the consequences of different instructions. A further problematic point consists in determining the right step size in the successive construction of an MCS, which means to be able to anticipate the desires of the user not as much as possible but as much as needed by the actual context, i. e. completing gradually and not at once what is left unspecified. A too far-reaching anticipation could in fact cause a lot of retractions to be done if the user intended to follow a different construction path (s. [dP91, p. 21f]) for further considerations).

1.3 Associative, memory-based reasoning

The EXCEPT system offers basic representation primitives (premises, assumptions and defaults) that can be incrementally constructed and retracted so as to serve as flexible building blocks for the specification of complex domain and user preference models. If we stick, nevertheless, to additional representation and inference techniques, so for many different reasons. A first general reason concerns the problem of handling global versus local inconsistency. In the logic-oriented framework inconsistency is a local phenomenon with global effects. In other words, having, for instance, two contradictory clauses in a set of clauses renders the whole set useless in the sense that everything can be deduced from it (this is the classical Aristotelian principle *ex contradictione sequitur quodlibet*). Such a situation can, in general, only be 'repaired' by isolating all the contradictory subsets, which is a potentially exponential time consuming task as one has to consider, in the limit, the whole power set of a given set of clauses[1]. Handling inconsistency in a frame-based representation is, in general, less costly. Frame structures do not stand for isolated but for highly interrelated chunks of knowledge, connected, for instance, through specialization (inheritance) links. Doing inferences consists in traversing paths along such links. As frames can be focused on individually, they provide flexible entry points into a net. This means that one seldom has to consider the whole frame network and, depending on the form of a subnet and the type of links involved [THT87], one can be sure to make, within it, correct inferences without having to care for other, possibly inconsistency bearing parts of the same network. This local safety is, of course, gained at the cost of a more limited inference capability. Traversing links is, on the other hand, much more efficient than doing resolution steps in a logic-oriented language. But more than efficiency, the basic point here is that the need for global consistency is reduced or, loosely speaking, it is taken the bite out of it.

---

[1] Note that computing *minimal (smallest) inconsistent* subsets is the inverse problem to the problem, mentioned above, of computing the *maximal consistent* subsets of a given set of clauses.

Other important reasons for trying alternative, frame-oriented forms of knowledge representation are related to the properties of the application field. The point to note here is that we have to cope with objects and situations that have an "extension" (can be represented graphically) and that are related to each other in the sense of part-whole relationships. It is generally agreed that handling complex compound objects, i. e. objects composed of parts, each part being either another complex or a primitive object, is not easy for logic-oriented representation languages like EXCEPT. They have, in this respect, the same problems as those known in the context of relational database languages. If required, a complex object can here only be represented by a multitude of clauses (tuples). In this way, the object is, as [Re89, p. 11] puts it, "absolutely non-existent" and can only be "(re-)assembled" from the different parts through appropriate user queries. That means that the responsibility for the reconstruction of a complex object lies with the user and not with the (database) system. A frame oriented representation does not suffer from this problem. Another point is that reasoning about extensions means, in the extreme case, to be capable of doing analytical geometry, which is too a complex (time consuming) task in a pure logic oriented framework. Our intention, in this respect, is even to dispense completely with analytical geometry and to study and apply, instead, what we call *model-based* methods of spatial representation and reasoning.

1.3.1 Model-based vs. syntactic reasoning

Because doing spatial or graphic reasoning in a model-based way is a basic concern for the project, we want to give here a short example, which illustrates our understanding of the expression "model-based" and which helps to clarify the underlying motivation. Model-based reasoning is best viewed in contrast to purely propositional (syntactic) reasoning. Example of propositional reasoning:

Consider two different sets (M1 and M2) of objects, where the elements of M2 are sets, which contain elements of M1. Suppose the following is known to hold:

• Any two members of M1 are contained in just one member of M2.
• No member of M1 is contained in more than two members of M2.
• The members of M1 are not all contained in a single member of M2.
• Any two members of M2 contain just one member of M1.
• No member of M2 contains more than two members of M1.

From this small set of premises we can derive a number of theorems by using customary rules of inference, such as they are implemented in systems like EXCEPT. It can be shown, for instance, that M1 contains just three elements. Is the syntactic transformation of premises by rules of inference the only way of making explicit what is implicit in them? Another possibility consists in building a geometrical model that gives a "concrete" spatial interpretation, i. e. a semantic, for the symbols (naming variables and relations) used in the premises. Take, for instance, the elements of M2 "to be" the sides, and the elements of M1 to be the vertices of a triangle:

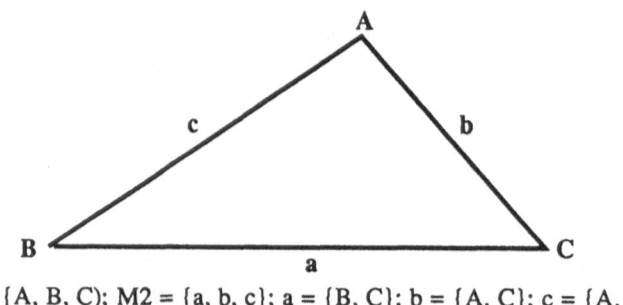

M1 = {A, B, C}; M2 = {a, b, c}; a = {B, C}; b = {A, C}; c = {A, B}

Then you "see" that the above theorem holds[1]. There is evidence to suppose that most of the reasoning human beings do (not only in common situations) is of a similar form, i. e. mental images (or, in general, models) play a decisive role [NC91]. They are, possibly, the original (in the (chrono)logical sense) way to tackle problems. As another, more everyday example, the reader is asked to think over what s/he will do when requested to give an answer to the following question: How many windows are there in the front side of the house you live in? Probably, s/he will build a (mental) image of the house front and then count the windows.

Mentals models have been carefully studied by [JL83] for finding answers to questions like 'What happens when we understand a sentence?' or 'How is it possible for you to make a valid deduction even if you have not learned logic?'. The same author summarizes in [JL88, p. 226f] the difference between propositional reasoning, i.e. working with formal rules that are purely syntactic, and model-based reasoning (working semantically on the basis of 'vivid images') as follows:

> One way in which a valid inference can be made is to imagine the situation described by the premises, then to formulate an informative conclusion which is true in that situation, and finally to consider if there is any way in which the conclusion could be false. ... To imagine a situation is to construct a 'mental model' ... based on the meaning of the premises, not their syntactic form, and on any general knowledge triggered by their interpretation.

---

[1] We have borrowed and adapted this example from [Na59, p. 16ff] where it is used with the different intention to show how models can help to establish the consistency of a set of premises.

Of course, such a model-based way of proceeding perfectly fits the memory-based approach. Our basic research goal is to devise computational mechanisms which 'mimic' such mental processes. In this respect, we believe that (high level) frame-based structures supplemented by (low level) depictions, as a mean of realizing spatial models [Ha89], are a good starting point in so far as they allow to handle knowledge of various levels and to choose different degrees of granularity for the representation.

1.3.2 The ASM system

A first concrete result toward the goals described above is the Associative Memory Model (ASM) of [He91], a flexible experimental system which tackles different problems at different levels. Its overall organization looks as follows:

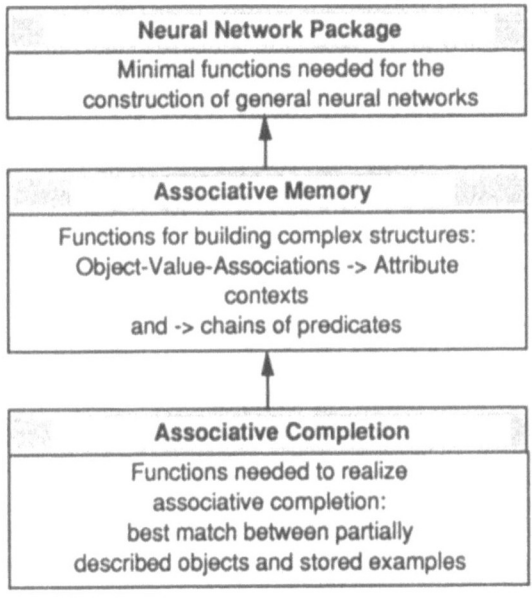

Structure of ASM

The basic level consists of a neural network package which realises the minimal functionalities needed to build general neural networks. The focus lies here on the requirement for minimality and the care not to be completely unrealistic from a biological point of view. The second level offers various forms of associative memory structures like object-attribute-value triples or chains of predicates. Functionally, the user has the possibility, for instance, to input triples and to retrieve them using attributes as context selecting devices. The important thing is that these structures are all based on the same primitive structures (nodes) and operated upon by the same basic "inference" method, that is a relaxing, value passing, spreading activation mechanism. Answers of the system are modeled as a network of mutually reinforcing nodes. At the third level, it is shown how these structures can be used for retrieval and associative completion. Previously stored (memorized) examples of objects and situations are retrieved on the basis of partial descriptions. Means for computing best matches (such as the intersection of attribute-values) are also realized using the same spreading activation mechanism.

A basic and still unresolved issue we are working on is the handling of incrementality with respect to the following questions. How do successive inputs merge in the memory and build abstract structures (that is the (old and well-known) problem of learning object classes from examples). What does "abstraction" at different levels mean, respectively, what does it actually mean to have abstraction in an extensional representation form. Supposing to have an answer to this problem, how can exceptions be handled. Furthermore, what is, in this framework, the meaning of negation, disjunctive information and inconsistency. As one can see, these are, besides the learning problems, almost the same questions which have been asked and answered in the context of non-monoting reasoning à la EXCEPT. The starting point is, of course, very different (only minimal conditions are assumed beforehand). The intention is different too, because the goal, here, is to remain as much as possible at the level of an explicit and extensional representation of knowledge.

In this respect, our view of associative, model-based reasoning joins aspects of research currently done in the context of so-called *vivid reasoning* [EBK89]. In the broader context of hybrid reasoning this approach contrasts with other attempts looking at limited inference or syntactic restrictions on the representation. The working hypothesis underlying this research is that some kinds of (fast) common sense reasoning are best modeled by simple database-style lookups. Starting from facts that may be presented in a more expressive (first-order) language, a vivid knowledge base (KB) is constructed such as to contain only elements (ground, atomic facts) which are in a one-to-one structural relationship to the parts of the world being modeled. In this way, a vivid KB becomes, like a picture, an *analogon* of the domain, i. e. everything that is represented is explicitly represented: "The notion of vivid representation is appealing for reasons beyond supporting reasoning as database-style lookup: it corresponds well to the kind of information expressed in pictures" [EBK89, p. 1147]. An interesting point, in this respect, is how incomplete, for instance, *disjunctive* information is 'vivified'. When a given information asserts that a particular individual has one of several properties, without specifying which (for instance, 'Joe is teacher *or* professor'), then a property is sought that subsumes the mentioned properties (this can be done with the help of a lattice of predicates providing subsumption information) and used to assert in the vivid KB a ground (atomic) sentence (for instance, 'Joe is instructor'). This amounts to substituting *vagueness* (the term 'instructor' is more general than (i. e. subsumes) 'teacher' and 'professor') for *ambiguity* (the disjunctive information) [EBK89, p. 1148 f]. Currently, the authors do not have a satisfactory treatment for the 'vivification' of *negative* information. A further idea in this context, which is also our (long standing) idea while trying to understand the transitions from associative to other forms of reasoning and the principles that make them possible, is that vivid (associative) reasoning has to supplement but not to replace a first-order one. In this sense, if the result of querying the vivid KB (the associative memory) yields insufficient information, one can then query the first-order (EXCEPT) KB, where more powerful but less efficient reasoning methods are used.

1.3.3 Autoassociative completion

A second result in associative reasoning tackles the problem of representing and handling spatial knowledge on the basis of neural network techniques [Pa92]. After discussing different forms of local and distributed representation, the author shows how to combine the ideas and intentions underlying object centered representations and feature maps. While feature maps are based on a fixed global coordinate system in space and consider the contents of each element of a discrete grid, object centered representations describe the locations of objects relatively, i. e. objects are described with respect to some variable, possibly nonuniform grid centered on another (reference) object. The problem with object centered representations is that one has to cope with a, normally, variable number of objects in the neighborhood of a given object. The apparent inflexibility of feature maps can, on the other hand, be (partly) remedied using overlapping receptive fields which are distributed over the map(s). If the receptive fields are big enough, they will cover most of the relevant relations between objects, i. e. those depending on their immediate neighborhood. This way, the associative procedures for the different fields may be identical, i. e. have the same parameters regardless of their location. The main advantage here is that fields of the same size are easily translated into an associative model with fixed input and output vector. The author shows how, according to this view, invariance with respect to shift and rotation may be achieved.

A possible extension of this approach consists in using several feature maps with different granularity or resolution and/or receptive fields of different size. This amounts then to using in the background several neural networks, where the (architectural) problem arises of how to compose (connect) them. A first step toward a solution could be to proceed along the lines suggested by [Fe87] (s. also [Pa92, p. 41ff]), i. e. to relate the maps to reference frames (like retina and head) having a different 'positional' value in the

110

context of a more complex and 'complete' cognitive system, that is a system which is not only able to 'see' but which is also able, for instance, to 'move' and which has 'knowledge'. The reference frame labeled 'world knowledge network' is indeed meant by [Fe87] to represent the agent's general knowledge of the world, including aspects independent of space or vision. This reference frame could be realized by or act, in our context, as an interface to other, higher level reasoning subsystems, like ASM or EXCEPT.

2. Examples of Interaction

At this point, we would like to give the reader a concrete feeling for the meaning of the assisting capabilities achieved so far. As application example, we take the specification of office furniture layouts. A concrete task, in this context, would be to select and arrange equipment for an office such that two office workers can do their work. This can be viewed as a complex design problem. The complexity lies in the necessity to consider different types of (quantitative, qualitative and spatial) information, as can be appreciated by the following short characterization of possible requirements:

- The office should have at least one telephone.
- The costs for the furniture should not exceed 6,000 DM.
- The desks are to be placed in such a way that the workers can directly see each other and that they enjoy an optimal lighting.
- Desks and filing-cabinets should be surrounded by sufficient free place.
- The distances between desks and filing-cabinets should be as short as possible.
- No object should block the way to a door or a window.

The requirements clearly range over very different, i. e. ergonomic, financial, positioning (spatial) and even social aspects. To the same degree as these can be given different weights or priorities, they can be used to build different preference models. In our opinion, it is of no (very intelligent) use to consider all these aspects together and at the same time so as to face a "huge", unstructured constraint problem one could then try to solve 'monolithically' with one optimization procedure. Instead, we want different subsystems (reasoners), that handle different aspects, to cooperate in solving the whole problem. Besides this point, that justifies our different methodological approaches and developments, the central problem, which the primary emphasis of the project is placed on, is how to cope with the inexact specification of such requirements (What is the meaning of "sufficient free place"? What does it mean to enjoy optimal lighting? Are there requirements that contradict each other? and so on).

We give now an example of our actual understanding and handling of these aspects in form of three snapshots of the interaction with an office layout assisting system. First, however, we present the architectural arrangement of the subsystems employed in form of a graphic figure in order to give the reader an orienting overview (s. next page).

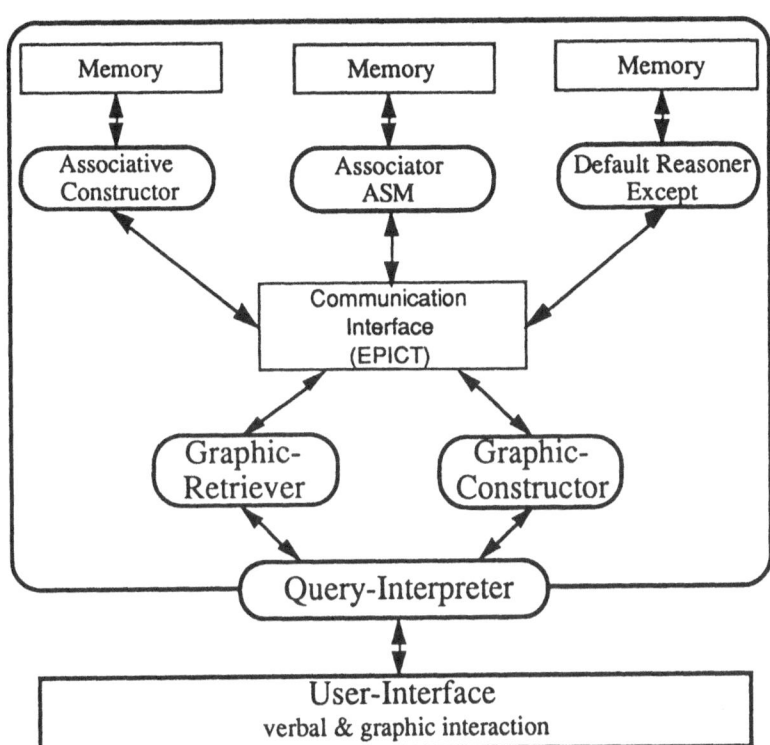

The different components are in rather different development stages. Most advanced is the development of the EXCEPT system, whose version II is available as fully documented free software. A unifying user interface, which allows mode switching (graphic search & construction) on the basis of a dialogue management system (query interpreter), is still missing. The language EPICT [RWB90], however, which is necessary not only for graphic representation purposes but also as a common communication mean among the internal components is almost complete.

2.1 Handling contradictory instructions

The first snapshot concerns the handling of contradictory requirements. The background reasoner used is actually not [dP91, p. 23f] but could be the EXCEPT system. Suppose the design of an office layout has progressed as shown in the following screen section:

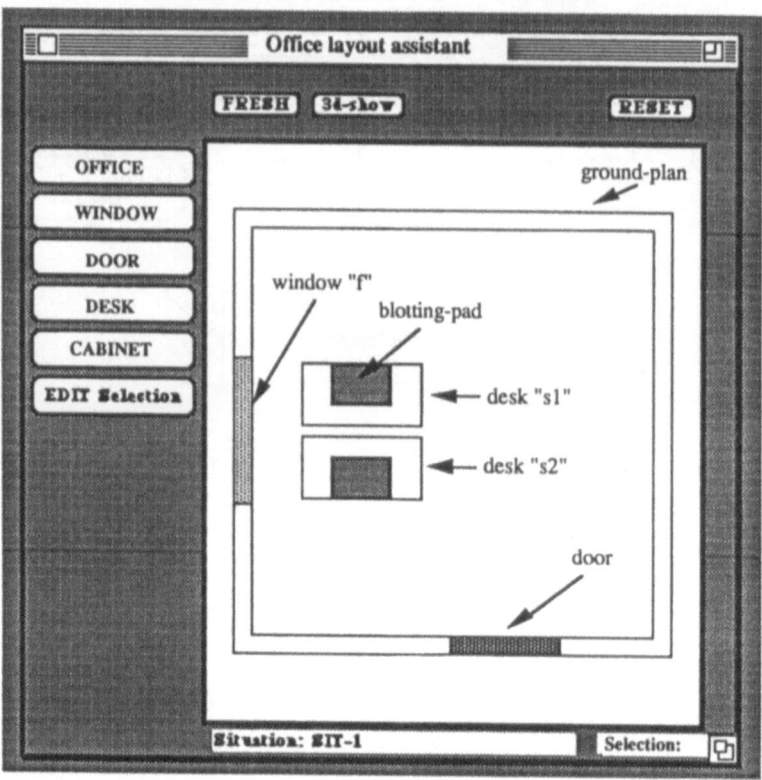

This is (excepted the words with outgoing arrows, which are annotations for a better understanding of the graphic symbols) the current interface of our office layout 'assistant'. Note the position of the blotting-pads, which indicate the side of a desk, where a worker would sit. Desks, windows and doors are objects which have an orientation in the sense of an intrinsic reference frame [RS88], i. e. they have (like people) a front, back, right and left side (in the two-dimensional representation upper and bottom side are not taken into account). In this situation, the user requests the system to change the layout in such a way that some (further) requirements are fulfilled. The user input (preceded by the prompt "C <-:", where C means "construct") is written in bold-face type.

C <- :   ((vis-a-vis s1 s2))
           ((optimal-lighting f s1))
           ((optimal-lighting f s2))
           ((for-right-hander s1))
           ((for-right-hander s2))

These are five requirements, that cannot (as the system will find out) be fulfilled at the same time. The first one means that the desks are to be positioned such that the workers can directly see each other (sit vis-a-vis). The second and third one mean that the desks s1 and s2 should receive an optimal lighting from the window f. The last two requirements transmit to the system the information that the users of the desks are both right-handers. Note that this is a purely declarative form of instruction specification. The user only states which properties the desired result is required to exhibit but does not say (and not know) how the desired result is to be obtained[1]. The answer of the system (marked with "A -> :") is:

A -> :   In the current situation, one of the five requirements ((vis-a-vis s1 s2)) is already fulfilled. The remaining four cannot be satisfied simultaneously. There are two subsets of them expressing properties which are mutually compatible:
           1: ((for-right-hander s2)) ((optimal-lighting f s2)) ((optimal-lighting f s1))
           2: ((for-right-hander s2)) ((for-right-hander s1)) ((optimal-lighting f s2))
           Please, choose (1 or 2) ?

Thus, the performance of the system is twofold. First, it has recognized that one of the requirements is already satisfied, i. e. it can be deduced from the premises (the current state of the situation). Second, it has found two maximal subsets (MCS) of the remaining requirements which can be used to extend the current situation without producing inconsistency. In order to be able to give such an answer, the system must, of course, have a lot of background knowledge and reasoning capabilities. It must know, for instance, that a desk which is to be used by a right-hander is not best positioned if there is only a window on its right side (for a right-hander, lighting is optimal only when coming from the left). Thus, placing two desks (for two right-handers in an office with only one window) vis-a-vis is not compatible with having them both in optimal lighting position. Further aspects of the knowledge required in this context is given in [Ju91, p. 85ff] and [dP91].

---

[1] This remark is also valid for the examples in the next sections. Imperative elements do, however, appear if the user directly 'intervenes' in the graphic representation moving, for instance, with the 'mouse' an object from one to another position. In this case, the result of this action can (must) be translated into a set of propositions to enable the logical reasoner to handle them correctly. The translation is not a trivial task.

2.2 Associative search of office layouts

The following figure illustrates the use of the ASM system for doing graphic search. The idea, here, is to do search on the basis of queries that are (possibly annotated) sketches of office layouts. The system is supposed to already know several concrete examples of office layouts. Its task is to find, among them, (if possible, exactly) one that matches as well as possible the search query, i. e. the sketch. The upper part of the figure shows on the left side a concrete office layout (labeled SIT1) consisting of two desks in front of each other (vis-a-vis) and a cabinet on the right side of the window. The right side of the upper part shows the (internal) mechanism which captures the geometrical information of the layout.

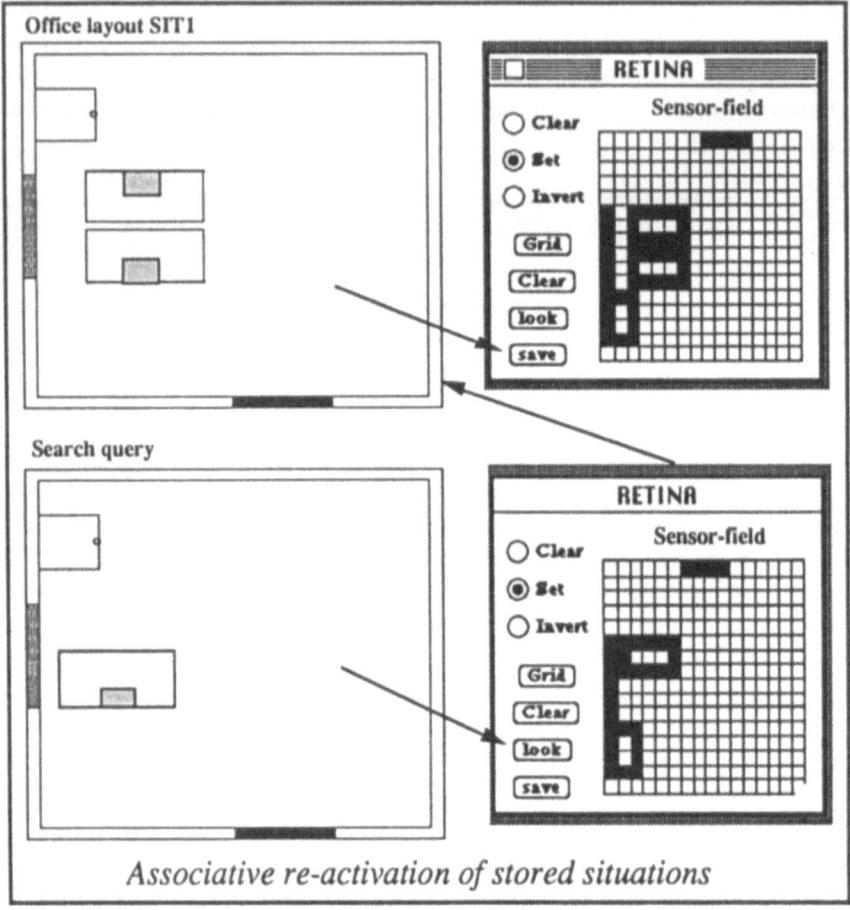

Associative re-activation of stored situations

This mechanism (which is not meant to be visible for an end user) is labeled RETINA in the figure and consists basically of a sensor-field realizing a simple pixel feature map. The granularity of the map is chosen to be 16x16 units but could be varied, realizing higher or lower forms of resolution. In this way, the exact information of the example is retained in the associative memory (after saving) only in a coarse form (note that the position of the objects in the map is reflected along the horizontal middle axis).

The pixels (i. e. the black fields of the map) are parts of the different objects. We say that they have different 'colors' depending on the 'type' of the objects they belong to. Thus, there are desk pixels, door pixels and so on. This type information is also saved away. Additional information, including aspects dealing with the general context in which the office layout has been produced, can also be stored. This means, after all, that a graphic situation can be annotated locally or globally with 'propositional' information. In this respect, the ASM system has proved to be very valuable because it can flexibly deal with different representations of both aspects, the geometrical (spatially organized) and non-geometrical features of the objects. The recall or re-activation phase, which is outlined in the lower part of the figure, is initiated (through the look-button) on the basis of a (possibly annotated) sketch of what the user is looking for. The query is content-oriented in the sense that, besides the basic graphic symbols, no particular syntactic restrictions are imposed for its specification.

Please note that the sketch is inexact in several respects. It is incomplete (only some objects are specified) and imprecise in the positioning of the objects included in the specifications. Part of this imprecision is handled (filtered out) at the level of the RETINA because of the coarse translation in the sensor-field (this can be seen as a sort of 'intelligent' preprocessor). The recall strategy is actually implemented in such a way that geometrical aspects, i. e. the information about the form, is considered first. If the result is insufficient (too many 'hits' or no hit at all), then additional information (type and further annotations) are taken into account. In the example shown in the above figure the search query yields exactly the office layout SIT1. We had here no big problem with multiple hits because, altogether, only about ten office layouts were stored in the associative memory.

2.3 Associative construction of office layouts

In this last section, we give an example illustrating the use of autoassociative techniques for the construction of office layouts. To this purpose, a more flexible form of interfacing is needed. We have to be able to do associative completion locally and not only glo-

116

bally. While for searching purposes it might suffice to partially specify a whole office layout and to get a (set of) known layout(s) as result, when constructing one wants, of course, to use previous examples but not to reproduce (copy) them in toto. One rather wants to adopt and adapt only parts of known layouts. In other words, a generalization capability is required. Technically, this means that one has to be able to look into the parts of existing office layouts and see and use similarities at a local level. This has been achieved using receptive fields (as explained in section 1.3.2). As before, the starting point is an adequate feature map where relevant information is encoded, for instance (the figures stem from [Pa92, p. 52]):

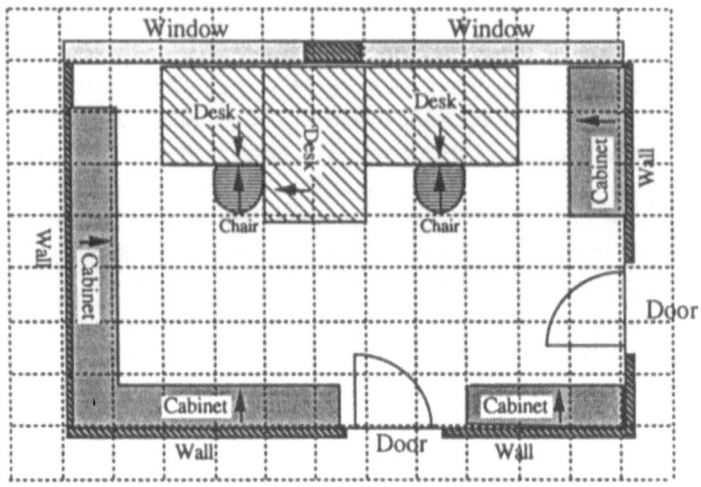

An Office with a Coarse Grid

Note that in the second figure (s. next page) showing the encoding of the above layout not only type information is represented but also information about the orientation of (the parts of) the objects. Thus, we have here not one but two (overlapping) feature maps. In the learning or training phase, which operates on the basis of specifications (examples) like the above one or parts of them, receptive fields (of, say, 3x3 size) are built which cover the whole feature map(s). The fields may overlap and have different orientation. Instead of using many different fields in parallel, it is, of course, possible to have one receptive field scanning the map(s) sequentially. The important thing is that the contents of the field(s) are translated in the same associative network. The overlapping of fields results in an invariance with respect to shift. Different orientations are needed to achieve invariance with respect to rotation (the use of 9 square pixels per field allows, of course, only a few useful rotations). In the recall phase, partial layouts or parts (for instance, corners) of layouts can be used which are different (in size and contents) from those used for training.

En	Fn	Fn	Fn	Fn	Fn	Wn	Fn	Fn	Fn	Fn	Fn	Ee
Ww		Ss	Ss	Sw	Sw	Ss	Ss	Ss			Rw	We
Ww	Re	Ss	Ss	Sw	Sw	Ss	Ss	Ss			Rw	We
Ww	Re		Cn	Sw	Sw		Cn				Rw	We
Ww	Re											We
Ww	Re											De
Ww	Re											De
Ww	Re	Rn	Rn	Rn	Rn	Rn			Rn	Rn	Rn	We
Ew	Ws	Ws	Ws	Ws	Ws	Ws	Ds	Ds	Ws	Ws	Ws	Es

p.of: part of a

C: Chair	D: p.of Door	E: Corner	F: p.of Window
R: p.of Cabinet	S: p.of Desk	W: p.of Wall	-: Space
n: north	w: west	s: south	e: east

Active Units of Feature Maps Encoding the above Office

A 'construction' query like the following one (where the stars indicate missing information):

```
En Wn Fn Fn Fn Wn Fn Fn Ee
Ww  *  *  *  *  *  *  * We
Ww  *  *  *  *  *  *  * We
Ww  *  *  *  *  *  *  * De
Ww  *  *  *  *  *  *  * De
Ww  *  *  *  *  *  *  * We
Ww  *  *  *  *  *  *  * We
Ew Ws Ws Ds Ws Fs Fs Ws Es
```

can, for instance, yield the following completions (the minus signs indicate free places):

118

```
En Wn Fn Fn Fn Wn Fn Fn Ee
Ww Ss Ss Ss Ss Ss Ss Ss We
Ww Re Cn -  Cn -  Cn -  We
Ww Re -  -  -  -  -  -  De
Ww Re -  -  -  -  -  Rw We
Ww -  -  -  -  -  -  Rw We
Ww Re Rn Rn Rn Sn Sn Sn We
Ew Ws Ws Ws Ws Fs Fs Ws Es


En Wn Fn Fn Fn Wn Fn Fn Ee
Ww Ss Ss Ss Ss Ss Ss Ss We
Ww Re -  -  Cn -  Cn -  We
Ww Re -  -  -  -  -  -  De
Ww Re -  -  -  -  -  -  We
Ww -  -  -  -  -  -  Rw We
Ww Rn Rn Rn Rn Sn Sn Sn We
Ew Ws Ws Ws Ws Fs Fs Ws Es
```

A substantial problem faced here is how to allow the user to give the system instructions containing negative, disjunctive and/or relational elements. An example would be: 'Construct an office layout with no cabinets' or 'Place on the west wall two cabinets with no other object between them' or 'Place a cabinet on the left or right side of a desk but not behind or in front of it'. Of course, it is difficult to represent such requirements, that express information of global or relational nature, in form of properties of single units of a feature map. Logically, a feature map corresponds (at best) to a monadic first-order language, i. e. a language which only allows the use of one-place predicate symbols. With respect to descriptive power, monadic predicate logic is of course weaker than general (n-place or polyadic) predicate logic (it is, however, decidable while first-order logic is only semi-decidable [Ha83, p. 123ff]). Even if, in some cases, the internal handling of such complex sentences could be easily done on the basis of units representing sum variables [Pa92, p.60f], it is unclear how to map them (supposing the user is allowed to formulate as many as s/he likes) onto fixed input vectors. Perhaps, the only practicable solution in this as well as in the context of ASM-based search, where the same problem arises, is to handle such aspects outside the associative network in a post-processing mode on the basis of other tools like EXCEPT. Thus, once more, the need for an hybrid approach to the whole problem becomes evident.

Acknowledgements

As leader of the TASSO project, I am expected (among other things) to give overviews stressing the relationships between different research aspects which are worked out by other people. In this respect, I want to thank particularly my colleagues P. Henne, U. Junker and G. Paaß, who have done the 'real' scientific work addressed in this paper.

References

[BA91]   W. Bechtel, A. Abrahamsen: "Connectionism and the Mind", Basil Blackwell, Cambridge (MA), 1991

[Bi89]   N. Bischof: "Ordnung und Organisation als heuristische Prinzipien des reduktiven Denkens", in: H. Meier (edt): Die Herausforderung der Evolutionsbiologie, Piper, München 1989, 79-127

[Br91]   G. Brewka: "Nonmonotonic Reasoning - Logical Foundations of Commonsense", Cambridge University Press, Cambridge, 1991

[CM84]   W. Clocksin, C.S. Mellish: "Programming in Prolog", Springer-Verlag 1984

[Con91]   L. Console, D. T. Dupre, P. Torasso, "On the Relationship Between Abduction and Deduction", in: Journal of Logic Computation, Vol. 1 No. 5, pp. 661-690, 1991

[DG88]   E. R.Dougherty, C. R. Giardina: "Mathematical Methods for Artificial Intelligence and Autonomous Systems", Prentice-Hall, 1988

[EBK89]   D. Etherington, A. Borgida, R. J. Brachman, H. Kautz, "Vivid knowledge and tractable reasoning: Preliminary report", in Proc. of IJCAI 89, 1146-1152

[Fe87]   J. A. Feldman: "A Functional Model of Vision and Space", in: M. A. Arbib, A. R. Hanson(Eds): Vision, Brain, and Cooperative Computation, MIT Press, 1987, 531-562

[Ha89]   Ch. Habel: "Propositional and depictorial representations of spatial knowledge: The case of *path*-concepts", in: R. Studer (edt): Natural Language and Logic, Springer-Verlag, 1989

120

[Ha83]   Ch. Habel: "Logische Systeme und Repräsentationsprobleme", in: B. Neumann (edt): "GWAI-83", Springer-Verlag, 1983

[He91]   P. Henne: "Ein experimentelles Assoziativspeicher-Modell", TASSO-Report Nr. 12, GMD St. Augustin, November 1990.

[Her91]   J.. Hertzberg: "Revising Planning Goals", TASSO-Report Nr. 3, GMD St. Augustin, March 1990.

[Ho91]   P. Hoschka: "Assisting Computer - A New Generation of Support Systems", in: Proc. 4. Internationaler GI-Kongreß "Wissensbasierte Systeme - Verteilte Künstliche Intelligenz", Springer-Verlag, Berlin, 1991

[JL88]   P. N. Johnson-Laird: "The Computer and the Mind", Fontana Press, 1988

[JL83]   P. N. Johnson-Laird: "Mental Models - Towards a Cognitive Science of Language, Inference, and Consciousness", Harvard University Press, Cambridge (MA), 1983

[Ju91]   U. Junker: "The EXCEPT II Default Reasoning System", TASSO-Report Nr. 23, GMD Sankt Augustin, 1991

[Mi80]   M. Minsky: "A Framework for Representing Knowledge", in: D. Metzing, (edt), Frame Conception and Text Understanding, de Gruyter 1980, 1-25

[Na59]   E. Nagel, J. R. Newman: "Gödel's Proof", Routledge & Kegan Paul LTD, London 1959

[NC91]   N. H. Narayanan, B. Chandrasekaran: "Reasoning Visually about Spatial Interactions", in: Proc. of IJCAI-91, 360-365

[Pa92]   G. Paass: "Representation of Multiple Objects for Associative Geometric Reasoning", TASSO-Report Nr. 35, GDM St. Augustin, January 1992

[dP91]   F. di Primio: "Verarbeitung ungenauer Anweisungen anhand des Beispiels Raumeinrichtung", TASSO-Report Nr. 26, GDM St. Augustin, August 1991

[Re89]   U. Reimer: "FRM: Ein Frame-Repräsentationsmodell und seine formale Semantik", Springer-Verlag, Berlin 1989

[RS88]   G. Retz-Schmidt: "Various Views on Spatial Prepositions", in: AI Magazine, Summer 1988, 95-105

[RWB90]   E. Rome, K.-H. Wittur, D. Bolz: "EPICT - Eine erweiterbare Grafik-Beschreibungssprache", TASSO-Report Nr. 4, GMD St. Augustin, Mai 1990

[Tha88]   A. Thaise (edt): "From Standard Logic to Logic Programming", John Wiley & Sons, Chichester, 1988

[THT87]   D.S. Touretzky, J. F. Horty, R. H.Thomason: "A Clash of Intuitions: The Current State of Nonmonotonic Multiple Inheritance Systems", in: Proc. of IJCAI 87, 476-482

Intelligent Interfaces
between Paper and Computer

Andreas Dengel, Rainer Hoch
German Research Center for Artificial Intelligence (DFKI)

Abstract

Many companies rely on converting existing printed material as well as incoming mail into an electronic representation that allows for automatic information management including content-based retrieval and distribution. Recent publications prove that OCR systems are to weak for such a task because they are adapted for data conversion rather than on a supply of information. Thus, there is a pressing need for document analysis and understanding systems that should be used as intelligent interfaces between paper and computer. This paper gives a survey on our research efforts in the ALV project — a German acronym for *Automatic Reading and Understanding* — and gives some ideas about possible applications.

1 Introduction

For many decades, paper has been the most important medium for indirect communication and storage of information in the office. In the 70th and early 80th, after computer technology had increasingly enhanced the capabilities for information interchange, many people had predicted the office in which paper would be obsolete. Carefully investigating and comparing strengths and weaknesses of both paper and electronic medium does not lead to any favorization of one medium [YMD85]. Rather, a co-existence of both information carriers in the foreseeable future will dominate our daily work in the office [Lev88].

Today, however, paper still is and seems to remain a very popular and unreplaceable information carrier. Although there are few examples of paperless offices for an inhouse interchange of information (ref. [Hou89]), paper will remain the standard medium for external communication.

However, the spread of computer technology not only has influenced the quality of information interchange, but also the quantity of information being produced. As a supplementary effect, paper consumption increases more than 10% every year [SF86]. Many studies and forecasts of institutions, like Battelle or Control Data Institute, confirm the dominance of paper. For example, in 1990 about 13% of all office costs were caused by consumption of paper and 90% of office time was filled with paper handling.

To limit and moreover to reduce paper consumption caused by rapid strides in publishing and printing technologies requires the development of interfaces for transforming printed information into symbolic representation. As result, it will be possible to keep originally printed information uniformly with other electronic data in one and the same electronic archive. Thus, the most important question is: how to get real-world documents into computer systems?

For transferring information from paper to computer, the most obvious way is using a scanner in combination with an optical character recognition system (OCR). Most of these systems are restricted to a small set of different fonts for which they first have to be trained during a preparatory phase. However, these systems are far away from a performance allowing an input of a printed sheet of paper and the system comes up with a coded file comparable in usability to the keyed-in version [CN91; Nag92; Bay92]. For instance, reading order is completely lost, paragraphs in documents having multiple columns are misordered, tables are garbled, headings extending a certain font size are not recognized, small fonts, such as in abstracts, are jammed by segmentation errors and misspellings.

Neglecting all of these aspects, even a perfect recognition of the characters in a printed text is only a data conversion between different media. This is insufficient for an integration of complex archival documents in electronic media and their consequent processing. Due to the amount of information available in printed form and to support humans in information filtering and management, tools should be provided acting as *intelligent interfaces* for the supply of information relevant for a user, a task, or a procedure.

To do so, it is not sufficient to traditionally pursue an isolated recognition of printed characters. Moreover, identification of the physical and the semantical structure of documents as well as determination of the type of information is required. Even in the domain of office information systems a new generation of sophisticated systems is required which help people to filter, sort and prioritize relevant information of documents. A few systems already fulfill such "intelligent" demands but are typically restricted and specialized to particular application areas such as electronic mail distribution and classification. Exemplary, one of the most famous systems is the *Information Lens* developed at MIT [MGT86]. The Information Lens is a knowledge-based system filtering relevant information of electronic mails exclusively. It exploits concepts from AI encompassing frames to represent message types, production rules and inheritance networks. A complete understanding of these messages involving natural language processing techniques, however, is not intended.

The ultimate goal of our activities — similar to the Information Lens — is not a complete interpretation of the document´s contents, but rather an identification of the central message conveyed. For example, in our domain of business letters, this includes recognition of the persons involved, such as *sender* and *recipient*, the type of message, such as *offer*, *order*, or *receipt*, but also aspects of information related to the message, such as *name of product, number* and *price*.

Consequently, our activities concentrate on the investigation of human reading techniques and the simulation and integration of various human knowledge sources for automatic capturing of printed information. For instance, such knowledge sources include expectations towards a text, geometric knowledge about the arrangement of text portions on a sheet of paper as well as syntactic and semantic knowledge.

124

This article gives an overview of our research in intelligent interfaces between paper and computers. In Section 2, we decribe our overall processing model, including four different processing steps: layout extraction, logical labeling, text recognition, and partial text analysis (ref. Figure 1). As example, Section 3 focusses on three application scenarios: inhouse mail or fax distribution, knowledge based indexing of printed documents as well as automatic task processing. Possible fields of applications are offices, libraries, assurances as well as postal and financial agencies, or other institutions in which a handling of structured documents is fundamental.

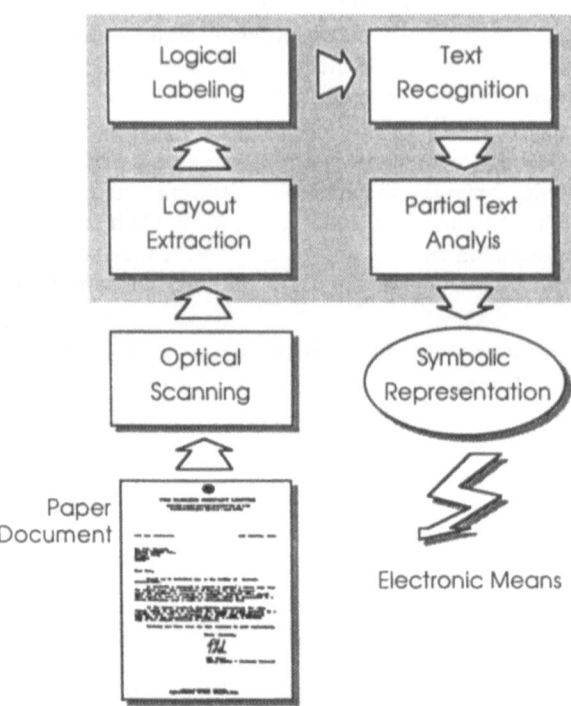

Figure 1: Overall processing model.

2 Processing Model

A document may be considered as a two-dimensional presentation of a collection of information that is adapted for human perception. The information may be of different modes like text, graphics as well as raster images, formulas, and tables. Such a document is not only characterized by its individual contents, but also by a logical organization into components that relate to a human perceptible meaning, e.g. the *author* of an article or the *recipient* of a letter. This logical structuring is done in order to enhance the comprehension of the contents. For visualization (displaying or printing), the information is formatted by defining corresponding two-dimensional presentation aspects, such as positions, shapes, and styles. The resulting layout structure, that is block order, line spacing, number of columns, etc., underlines the originator's intention of logical document organization.

Both, the layout structure and the logical structure are essential knowledge sources as well as an excellent orientation points for the analysis of document images. Therefore, the two initial steps of our system concentrate on structure analysis tasks.

First, an optical scanner transforms the printed information of a business letter into an electronic raster image. Subsequently, the document image is explored to extract a part-of hierarchy of nested layout objects, such as *blocks*, *words*, and *characters*, on the basis of the black-on-white presentation on the sheet (layout extraction). The layout objects and their compositions are geometrically analyzed to identify corresponding logical objects (e.g. *sender, recipient, date*). In this way, the layout structure is investigated for object arrangement by applying geometric rules which describe individual logical objects (logical labeling).

Further on, layout objects capturing word images are taken as input to generate a sequence of character hypotheses for a certain word position. Consequently, the character hypotheses are verified by checking them against dictionaries of valid words (text recognition). As result, text recognition yields a set of possible word candidates for higher-level analysis steps, i.e. document understanding.

During partial text analysis, those logical objects having a high degree of syntactic conventions are initially analyzed. To do so, grammars for each logical object, such as for *recipient, date*, or *subject*, are given as input to a syntactic parser. During this syntax check, respective logical objects are further refined. For example, the *recipient* is divided into *name* and *destination*, while the former, can be refined into *title, first name* and *last name, abbreviations*, etc. Moreover, syntactically analyzing text in logical objects allows an additional verification by refuting invalid word candidates.

Finally, the *letter body* as well as the *subject* of the letter are statistically examined with respect to word frequency. The occurrence and frequency of words as well as their positions in the text are evaluated according to specific message type keywords. The system then comes up with a set of weighted hypotheses about messages that might be conveyed in the given document, such as *order, invoice, confirmation*. This preclassification provides an expectation towards the contents of a letter being a starting point for applying sophisticated techniques for a semantic analysis.

Figure 1 shows the individual processing steps. In the following, we describe all analysis phases in more detail.

2.1 Layout Extraction

To get a first electronic representation of printed information, a document is scanned by a flatbed scanner at a resolution of 200 dpi providing an image of millions of isolated raster dots (pixels).

126

Layout extraction itself comprises several processing steps: skew angle detection and correction as well as image segmentation including image filtering, object detection, text/nontext discrimination, and object aggregation.

For locating text objects in the scanned document image, i.e. words and text lines, it is usually required that all text lines are along horizontal orientation. Therefore, if a document image is skewed, the skew angle has to be detected and consequently eliminated.

In literature, several well-known methods for skew detection and elimination are proposed [Bai87; DS89]. In our system, we prefer the so-called *Simulated Skew Scan* algorithm [Pos86]. This method is relatively expensive, but very reliable. It is mostly insensitive to noise and independent on font types and sizes. Although in some cases graphics may fool the detection, tests confirm that it is usually sufficient to detect angles from -4° to 4°.

After skew correction, the adjusted image is tracked for layout objects, such as blocks of information, and if appropriate refined to lines, etc. Consequently, the entire document can be represented by its layout structure. This function is often referred to as *document segmentation*.

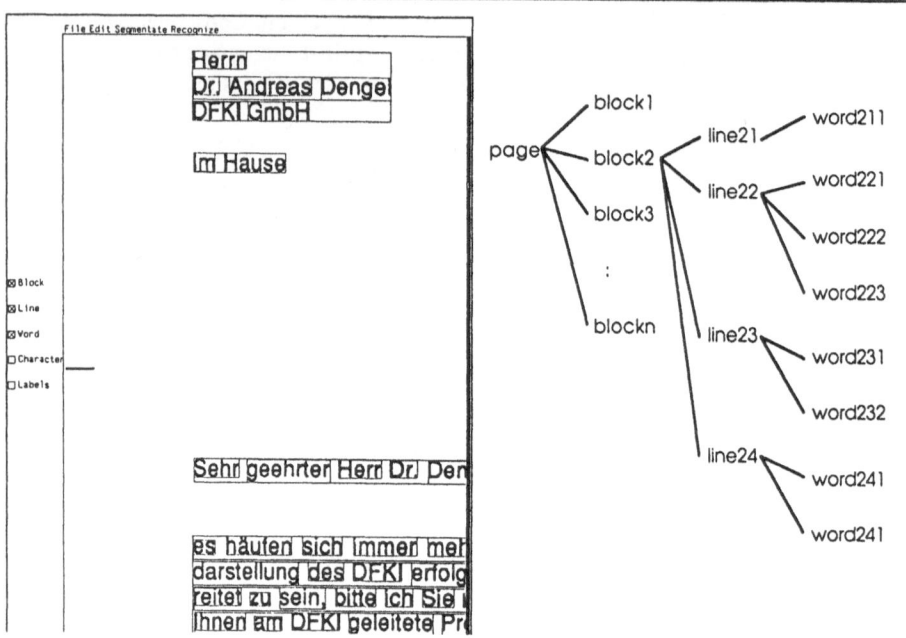

Figure 2: Exemplary result of layout extraction and layout structure representation.

We use a mixed top-down/bottom-up strategy for extracting the complete layout from blocks down to connected components. In a first step, paragraphs, lines, and words of the real document are stepwise extracted in a top-down manner. Subsequently, word images are tracked for connected component (of black pixels) describing characters and parts of them [Hön91; Den92a]. Here, we prefer methods on the basis of proximity, i.e. the *Run-Length Smoothing Algorithm* (RLSA) [WCW82] and interative projection profile cuts based on so-called *X-Y trees* [NS84; NSS86], gathering statistical data and distances for pixel groups.

A distinction between text and graphical objects is done by checking the size of connected pixel groups and their nesting structure. This is a weak classification, but gives us tolerable results, since we do not analyze graphics but rather focus on documents consisting mainly of text.

As output of layout extraction, information presentation is mapped into a tree-like structure capturing the document's contents and describing it geometrically. Figure 2 shows a clipping of a real layout extraction result as well as the respective symbolic representation in terms of hierarchically related layout objects.

2.2 Logical Labeling

The human's perception techniques during reading are mainly directed by experiences about the arrangement of information, i.e. text, on a document page. A nice example is to show a letter to a person from a couple of meters. Although she or he is not able to read any single character, the person may categorize the document as being a letter. Moreover, the person may associate meanings to document parts, such as *recipient*, *date*, *logo*, or *footer*. Hence, it seems useful to utilize this kind of knowledge source for attaching logical labels to layout objects designating human perceptible meaning and preliminary neclegtinging any access to the captured text [DB88].

Instead, the layout objects and the aggregates they build are taken as a geometric description of a given document. These are blocks of text and nontext, where the text blocks again consist of lines, etc., each of them being described by their positions and dimensions on the sheet.

For an attachment of logical labels, we distinguish between two geometric views (ref [Den92b]):

- A *global view* describing possible arrangements among logical objects.
- A *local view* describing geometric features of individual logical objects.

The knowledge necessary for logical labeling is composed of a so called *geometric tree* and a *statistical database* (SDB) [DB92b]. The tree is a specialization hierarchy, describing different arrangements of logical objects of business letters at different abstraction levels. The SDB provides isolated views to geometric features of individual logical objects represented in terms of rules, e.g.:

◊ the position of the recipient is in the upper third of the page;

◊ the left margin of the recipient is within the left quarter of the page;

Following a best-match heuristic [DB89], the given layout structure is explored with respect to the two geometric views. Applying a hypothesize & test strategy, a traversing of the *geometric tree* provides hypotheses about the logical meaning of

128

layout objects which consequently are verified by applying the corresponding set of geometric rules in the SDB. In this way, important meaningful parts of a document can be identified. Figure 3 shows results of logical labeling and schemes the internal mapping of the given layout structure into a respective logical structure.

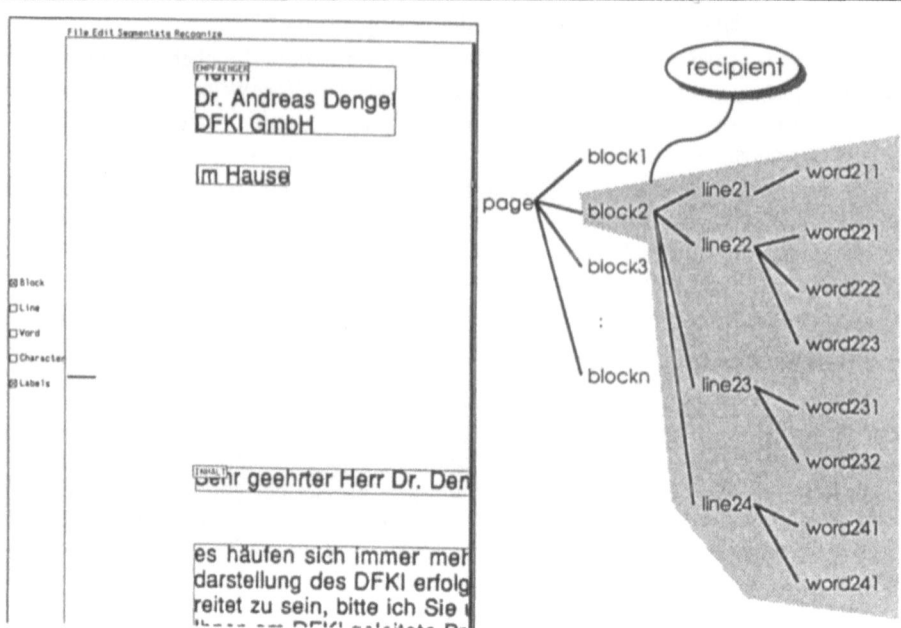

Figure 3: Result of logical labeling and mapping of layout and logical structure.

2.3 Text Recognition

The atomic units of text are the symbols of the alphabet. Beside the western lower and upper case letters and numerals, there are at least three dozens of punctuation symbols. Due to the progress in the fields of publishing and printing, there currently exists a wide variety of text fonts with their own styles. Different font types are designed by ligatures and kerning. Characters can follow each other by proportional or fixed distances. Different font styles, e.g. bold, italic, outline or underline, can occur in one and the same text block [Nag89; CN91]. Given this immense variety of type styles and forms in current use, it is not possible to assign alphabetic identities to characters of arbitrary size and typeface without additional knowledge beyond the entire image data level.

Based on this consideration, we prefer the word context for text recognition instead of a solely and isolated recognition of individual characters. This consideration reveals an important advantage: *words can be related to meanings as well as to con-*

texts. These, however, sometimes ambiguous meanings together with their interrelations within a phrase or a sentence enable a contextual recognition of text in logical objects and consequently provide an adequate basis for expectation-driven postordered text analysis.

For that purpose, we have implemented a text recognition specialist using word images decribed by respective layout objects as context for recognition. Within the given layout structure, words are further subdivided into layout objects capturing images of characters. These images are input for feature-based character recognizer which generates hypotheses for possible alphabetic identities. The features used are robust and limited in amount [Den92a]. As result, for each position of a word, character hypotheses are generated. Their combination provides a *character hypotheses lattice* (CHL) describing all possible concatenations of characters to strings.

In the context of words, the alternative strings can be verified using additional knowledge in form of dictionaries.

Because logical labeling provides the identification of meaningful document constituents, i.e. a restriction of context, character hypotheses verification can be performed in an expectation-driven manner. Consequently, the individual strings in the CHL are checked against lexical knowledge associated with specific logical objects representing legal words of an actual context (e.g. possible employee or city names) [DPH92]. Figure 4 illustrates an example of a CHL, the word candidates resulting from dictionary look-up as well as a clipping of practical results.

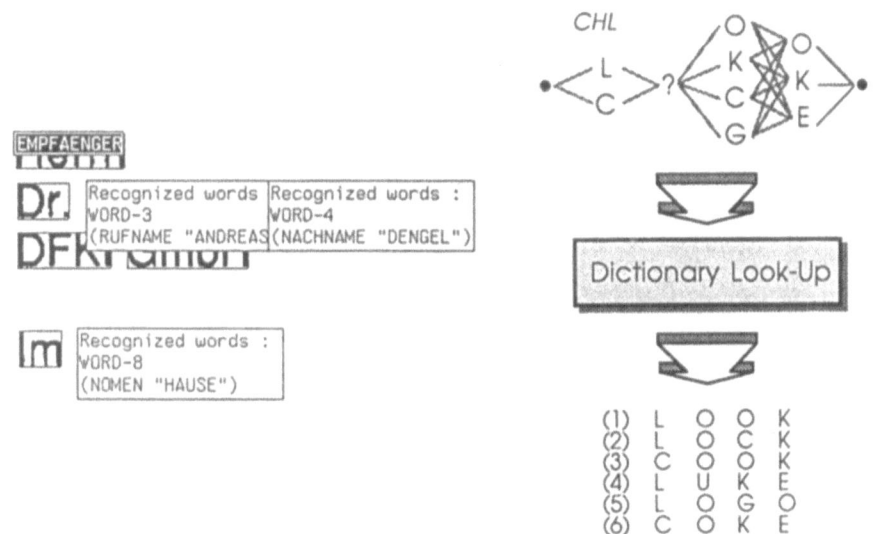

Figure 4: Scheme and results of dictionary look-up.

130

2.4 Partial Text Analysis

In our domain, each message type describes categories of information which a business letter may deal with, such as *offer, order, invoice* or *receipt*.

Each of these message types is a represented as a frame-like structure consisting of a collection of real world entities being related to each other. These entities may describe individuals, such as persons and animals, physical objects, abstract terms, or events which are characteristic for a certain message type. For example, an *offer* may be composed of entities *supplier, recipient, subject, price, number, date, etc.*. Addtionally, the entities may be hierarchically organized in an inheritance network (cf. Figure 5).

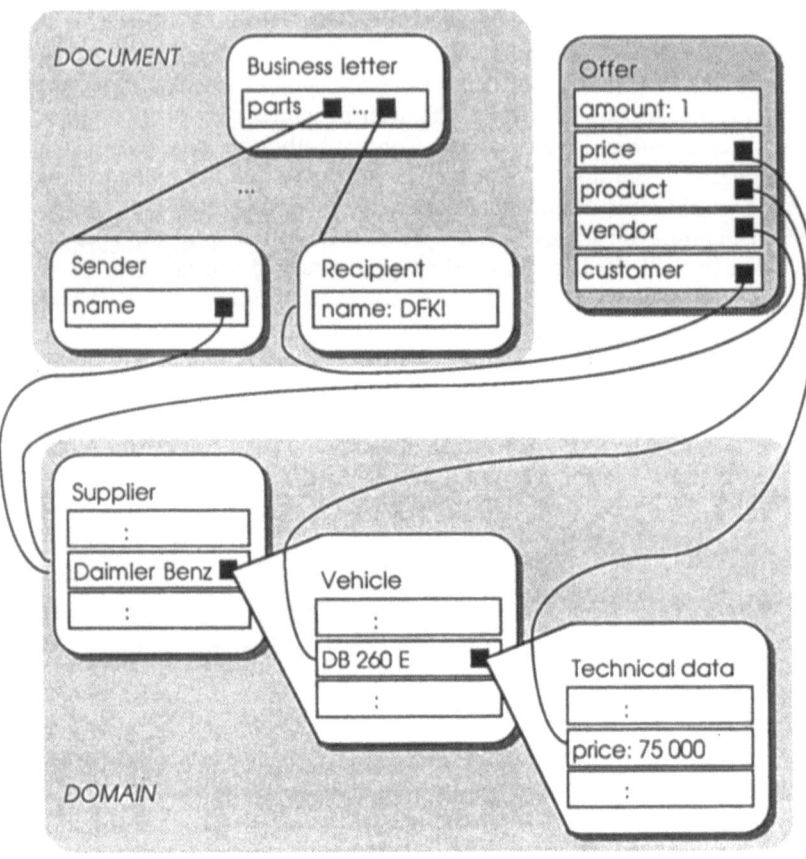

Figure 5: Example of message identification result.

For message identification, these entities have to be instantiated during text analysis. To achieve this goal, we initially use classical information retrieval techniques [Sal83, Sal91] to extract significant words of a letter on a purely statistical basis (index terms). These words are also weighted according to their relevance. Our experience shows that some words are most characteristic for particular message types indicating the class of a letter. For instance, typical offers include word inflections of the German verb "anbieten", "angeboten", "boten ... an" or respective synonyms.

It should be mentioned, however, that our indexing methods suffer from two inherent problems. On the one hand, our statistical database is not very large at moment (about 100 letters) implying a small set of modelled message types. On the other hand, typical business letters on their own are small comprising one or two pages mostly. Thus, classification results should be considered with caution.

In a next step, these — often ambiguous — classification hypotheses and weighted index terms form expectations towards the contents of a document. Actually, we evaluate several approaches of natural language processing (NLP) for filling the slots of our message type frames as well as establishing links between objects.

Primarily, we concentrate on skimming techniques such as implemented in the FRUMP [DeJ82] or SCISOR [RJ88] system. These systems accurately extract certain conceptual information from texts in preselected topic areas (e.g. news stories). Even the FRUMP system proved that an expectation-driven strategy was useful for interpreting texts in constrained domains. We belief that our domain of business letters and a corresponding message type model will allow similar skimming techniques for language analysis. In particular, our message types are comparable with the sketchy script idea presented in FRUMP.

Because the hypotheses about document semantics — yielded by our skimming component — can still be ambiguous like the ones gained from automatic indexing, it may become necessary to incorporate more NLP techniques. Thus, we also focus on powerful parsers in order to solve such ambiguities. In particular, we would like to integrate feature-based island parsing which can deal with incomplete recognition results as well as lexical word dependencies (e.g., verb and noun valency using a dependency grammar). Note that these language analysis modules are not implemented in our system up to now. Figure 5 shows an example of a final result like it is intended to be produced by our system

3 Applications

The result of our efforts should lead to a better integration of paper documents into a modern computerized office bridging the gap between paper and computer without loss of information.

The symbolic representation of an originally printed document in terms of structural, textual as well as semantic knowledge may be the basis for various postordered systems. Derivating structural information of a document, its meaningful and relevant parts as well as the contents, will enable an efficient processing by electronic means.

Application domains of these intelligent interfaces for document analysis are manyfold. In the rest of this section we will briefly illuminate three possible applications which are attractive to our opinion.

In large business organizations, plenty of letters arrive every day. Several clerks are occupied with sorting and distributing them. In order to automatically route this mail to the appropriate departments or employes, the captured information has to be recognized and understood at least partially in a context-sensitive way.

For mail distribution (see Figure 6), it may be sufficient to have a look at the *recipient*. Our system provides the identity by partially analyzing the corresponding part of the letter. Or, let us assume, that letters of a particular company have to be handled with priority. Partial analysis of the company logo provides the information required to place the document into the proper box.

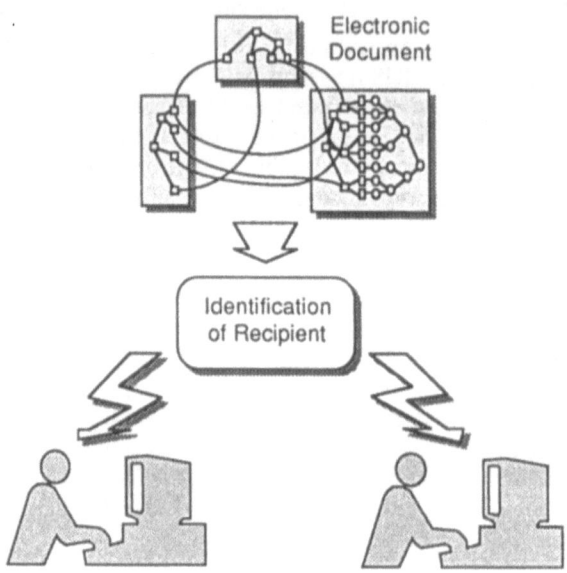

Figure 6: Automatic mail distribution.

This allows to send an incoming document to corresponding departments via electronic mail where it could be visualized on the screen. Consequently, the recipient himself is capable to decide whether to put it into garbage or to decide a further processing of the still partially iconic information. For instance, using OCR and serving a text or publishing system, it is possible to edit an individual document from the document database, or furthermore, to print or mail it.

At the same time, an intelligent multimedia filing system might incorporate all incoming documents in a document database which constitutes the central part of an office information system. Traditional document classification, especially keyword matching systems, representing the captured information by sets of index terms do not satisfy the requirements in offices [Sac84]. Consider the example in Figure 7, where a purchase department would like to retrieve all *offers* from a company *Siemens* being dated before *March 1st, 1991*. The system then must be able to differentiate between alternative message types, to know about a date and its semantics, and to realize that Siemens should be the sender of a letter.

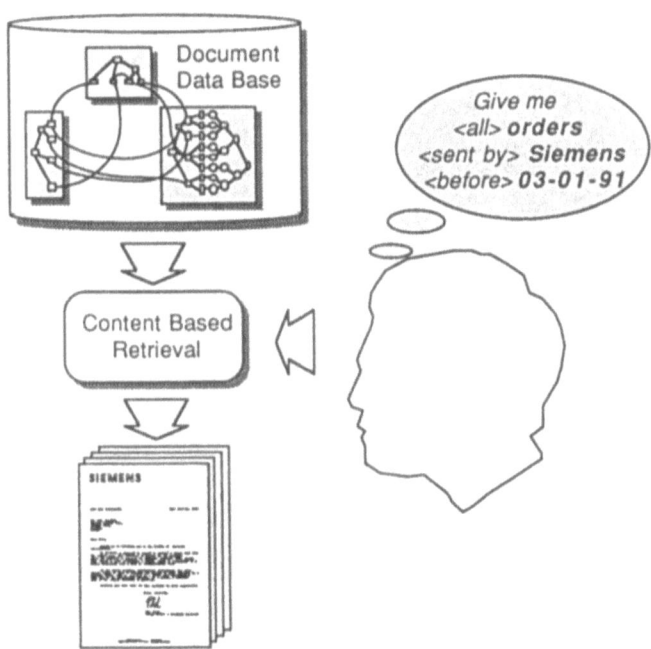

Figure 7: Filing and retrieval.

In this way, our system could be used as an interface to automatic content-based filing of documents. Thus, instead of huge paper archives, originally printed information may be manageable together with electronic data in one and the same document data base. Due to automatic document indexing, i.e. in terms of instantiated message types, users may be supported while searching for procedures, subjects, or events.

134

A more advanced and challenging application might be an automatic initiation of entire tasks. As a possible scenario, information being filtered out of an incoming document may notify different instances being related to a certain message and thus, initiates entire procedures (ref. Figure 8). For example, an order forces a request to the depot manager for sufficient stock and at the same time notifies the manufacture about the ordered product and the amount to be produced.

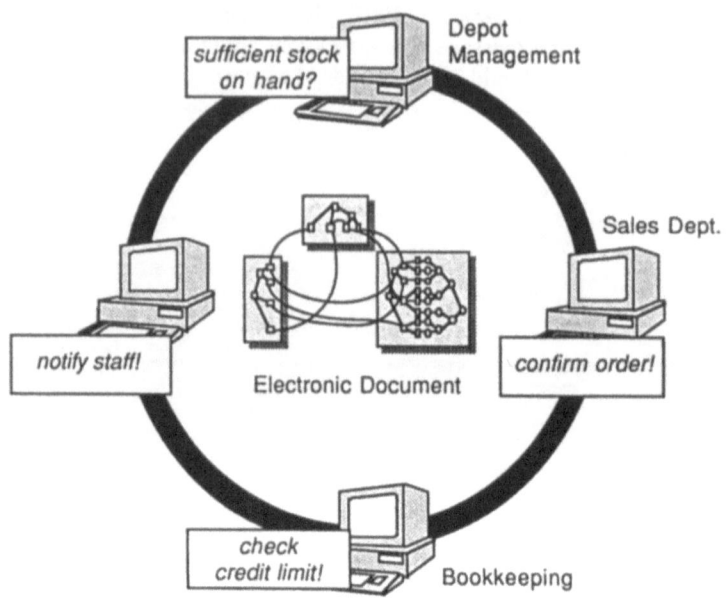

Figure 8: Automatic task processing.

These applications are small snapshots of the potential of intelligent interfaces from printed to symbolic information. They should, like the whole paper, serve as an introduction to the importance of the respective research field and the needs present in our daily life.

Acknowledgements

Our grateful thanks go to the German Ministry for Research and Technology (BMFT) which has been supported this work under contract ITW 9003 0. We would also like to thank our colleagues in the ALV team, DFKI, in particular Michael Malburg for detailed reading and valuable comments.

References

[Bai87]   H.S. Baird, " The Skew Angle in Printed Documents", Proceedings SPSE 40th Conference and Symposium on Hybrid Imaging Systems, Rochester, NY (1987), 21-24.

[Bay92]   T. Bayer, J. Franke, U. Kressel, E. Mandler, M. Oberländer and J. Schürmann, "Towards the Understanding of Printed Documents", in: H. Baird, H. Bunke, K. Yamamoto (eds.) Structured Document Image Analysis, Springer Publ. (1992), (accepted for publication).

[CN91]   R.C. Casey and G. Nagy, "Document Analysis — A Broader View", Proceedings ICDAR'91, St. Malo, France (1991), 839-849.

[DB88]   A. Dengel and G. Barth, "High Level Document Analysis Guided by Geometric Aspects", Internat. Journal on Pattern Recognition and AI, Vol. 2, Nr. 4 (1988), 641-656.

[DB89]   A. Dengel and G. Barth, "ANASTASIL: A Hybrid Knowledge-based System for Document Layout Analysis", Proceedings 11th IJCAI, Detroit, MI (1989), 1249-1254.

[Den92a]   A. Dengel, R. Bleisinger, R. Hoch, F. Fein, F. Hönes and M. Malburg, "ΠODA — The Paper Interface to ODA", DFKI Research Report RR-92-02, German Research Center for Artificial Intelligence (DFKI), Kaiserslautern, Germany (1992), 53 pages.

[Den92b]   A. Dengel, "ANASTASIL: A System for Low-Level and High-Level Geometric Analysis of Printed Documents", in: H. Baird, H. Bunke, K. Yamamoto (eds.) Structured Document Image Analysis, Springer Publ. (1992), (accepted for publication).

[DeJ82]   G. DeJong. An Overview of the FRUMP System. In W. G. Lehnert, M. H. Ringle (eds.), *Strategies for Natural Language Processing*. Lawrence Erlbaum Associates, Hillsdale, 1982, pp. 149-175.

[DPH92]   A. Dengel, A. Pleyer and R. Hoch, "Fragmentary String Matching by Selective Access to Hybrid Tries, Proceedings ICPR-92, Int'l Conference on Pattern Recognition, The Hague, The Netherlands (1992), (accepted for publication)

[DS89]   A. Dengel and E. Schweizer, "Rotationswinkelbestimmung in abgetasteten Dokumentbildern", Proceedings 11th DAGM-Symposium, Hamburg, Germany (1989), 274-278.

[Hou89]   D. Hough, "The Paperless Office", BYTE (July 1989) 241-246.

[Hön91]   F. Hönes, E.-G. Haffner, F. Fein and A. Dengel, "A Hybrid Approach for Document Image Segmentation and Encoding", Proceedings ICDAR'91, St. Malo, France (1991), 444-453.

[Lev88]   D. Leven, "Bridging the Gap between Paper and Electronics — The Paper Interface", iesnews, Esprit Information Exchange Systems, Issue No. 16 (June 1988), 15-17.

136

[MGT86] T. Malone, K. Grant and F. Turbak. "The Information Lens: An Intelligent *System for Information Sharing in Organizations*", *Proceedings CHI´86 Conference., Boston, MA (1986), 1-8.*

[Nag89] G. Nagy, "Document Analysis and Optical Character Recognition", Proceedings 5th Int´l Conference on Image Analysi and Processing, Positano, Italy (1989), 511-529.

[Nag92] G. Nagy, "At The Frontiers of OCR", IEEE Proceedings (to appear).

[NS84] G. Nagy and S. Seth, "Hierarchical Representation of Optically Scanned Documents", Proceedings 7th ICPR, Montreal (1984), 347.

[NS84] G. Nagy, S. Seth and S. Stoddard, "Document Analysis with an Expert System", Pattern Recognition in Practice II, Elsevier Science Publishers B.V (1986), 149-155.

[Pos86] W. Postl, "Detection of Linear Oblique Structures and Skew Scan in Digitized Documents", Proceedings 8th Int´l. Conference on Pattern Recognition, Paris, France (1986), 240.

[RJ88] L. F. Rau, P. S. Jacobs. Integrating top-down and bottom-up strategies in a text processing system. Proceedings 2nd Conference on Applied NLP, Austin, Texas (1988), 129-135.

[Sac84] G.M. Sacco, "OTTER — An Information Retrieval System for Office Automation", Proceedings 2nd ACM SIGOA Conference on Office Automation Systems, Toronto (1984).

[Sal83] G. Salton and M.J. McGill. "Introduction to Modern Information Retrieval", McGraw-Hill Computer Science Series, McGraw-Hill, Inc. (1983).

[Sal91] G. Salton and C. Buckley. "Developments in Automatic Text Retrieval", Science, vol. 253 (1991), 974-980.

[SF86] M. Schäfer and H.P. Fröschle, "Die Vision vom papierlosen Büro", Funkschau 19 (1986), 46-54.

[WCW82] K.Y. Wong, R.G. Casey and F.M. Wahl, "Document Analysis System", IBM J.Res.Dev., 26 (6) (1982).

[YMD85] N. Yankelovich, N. Meyrowitz and A. Van Dam, "Reading and Writing the Electronic Book", IEEE Computer (October 1985), 15-30.

Integration of Speech into Workstation-Based Applications

Peter Witschital, Ulrike Harke,
Meinrad Niemöller, Horst Schukat

Siemens Corporate Research and Development
Systems Technologies
System Ergonomics and Interaction
ZFE ST SN 7
Otto-Hahn-Ring 6
D-W8000 München 83
Federal Republic of Germany
witschi@zfe.siemens.de

Abstract

In contrast to human-computer interaction, people communicate essentially via speech. Moreover, humans are able to receive and understand visual information in parallel and synchronized with acoustic signals. Integration of speech into workstation-based applications promises an improvement of human-computer interaction towards a more natural interaction.

This paper deals with two different possibilities of integrating speech into workstation-based applications: voice annotation and speech command recognition.

Two prototypical applications are described which have been developed at the Siemens Corporate Research and Development Labs. The first one is a voice annotation editor which allows to integrate voice annotations into documents. In this case the computer is used as a medium which transports the speech information from one user to another.

The second example is a computer animation system which the user can operate by spoken commands. In this case the computer understands the words spoken by the user and interprets them as commands. Real-time speech command recognition improves the interaction with the system because the user can concentrate on the object of interest in the 3D-space.

138

1 Introduction

In communication between human beings not only the sense of sight but also the sense of hearing are used intensively (see Figure 1). Speech plays the major role in human-human communication. The sense of smell, the sense of taste, and the sense of touch are less important in communication and therefore not considered here.

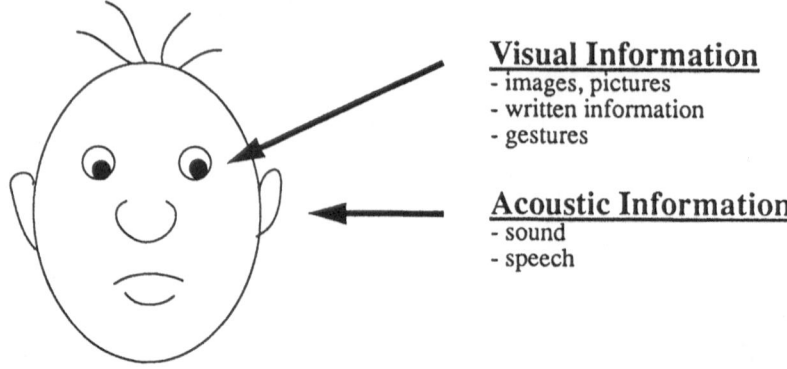

Figure 1. Human-Human Communication

In traditional human-computer interaction, however, the computer presents only visual information to the user and the user passes information to the computer by means of a keyboard or with pointing devices, like the mouse. The acoustic information channel is not used at all.

Because of the major role of speech in human communication it is only natural to integrate speech in- and output facilities into workstations in order to improve human-computer interaction and also in order to improve that part of human-human communication which is done by exchanging electronic documents between humans.

2 Voice Annotation

The first type of speech integration which is described in this paper is called voice annotation. Voice annotation means that there is a software tool which allows to record acoustic information in a way similar to a cassette recorder. The recorded information can then be imported into electronic multimedia documents and can be replayed at a later point of time (see Figure 2).

There are many applications where voice annotation can be useful. In electronic mail, e.g., spoken messages which have a much more personal character than written text can be added to the visual information. In computer based training applications spoken explanations or recorded sounds can be used to improve the instructional process.

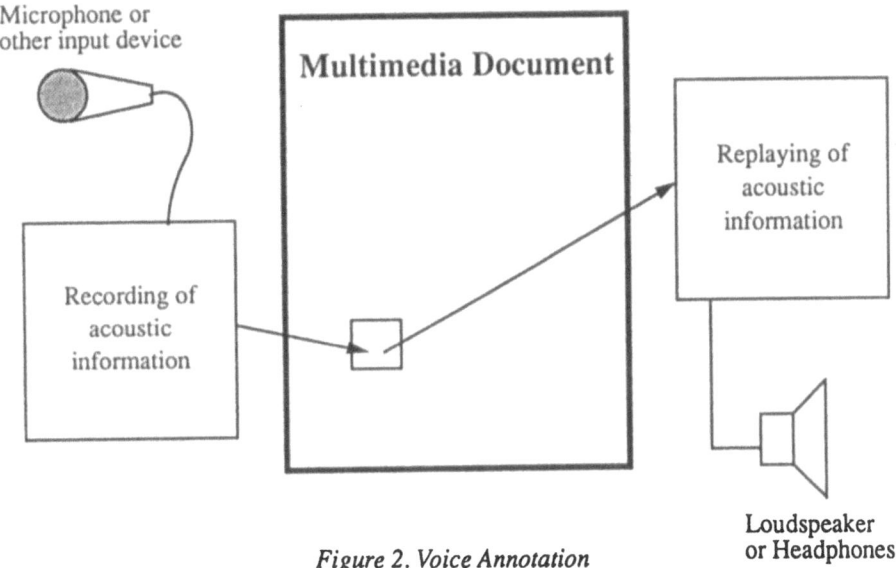

Figure 2. Voice Annotation

In a joint project between Siemens Nixdorf and the Siemens Corporate R&D Labs a voice annotation editor was developed which allows to integrate voice annotations into Frame-Maker[1] documents (see Figure 3). FrameMaker is a desktop publishing editor which allows not only to import graphics but also to work with so-called *insets*. An inset is a graphical representation created with a specially modified application called an inset editor [FM90]. An inset editor must be able to save graphics in a format FrameMaker can import. If an inset is imported into a document FrameMaker stores the name and location of the inset editor together with the imported graphics. The user can start the inset editor by double-clicking the graphical representation of the inset, change the inset, and then display the updated inset in a FrameMaker document.

To interact with other applications FrameMaker uses SUN's remote procedure call (RPC) interprocess communication standard [FM90.1]. SUN's RPC is a public domain communication protocol and is available on most UNIX[2] systems. The voice annotation editor was developed as an inset editor. The graphics which are imported and displayed in a Frame-Maker document are small loudspeaker icons, but together with the graphical information audio information is imported. If the voice annotation editor is invoked not the graphical information, i.e. the loudspeaker icon, is edited but the audio information.

---

1. FrameMaker is a registered trademark of Frame Technology Corporation
2. UNIX is a trademark of AT&T Bell Laboratories

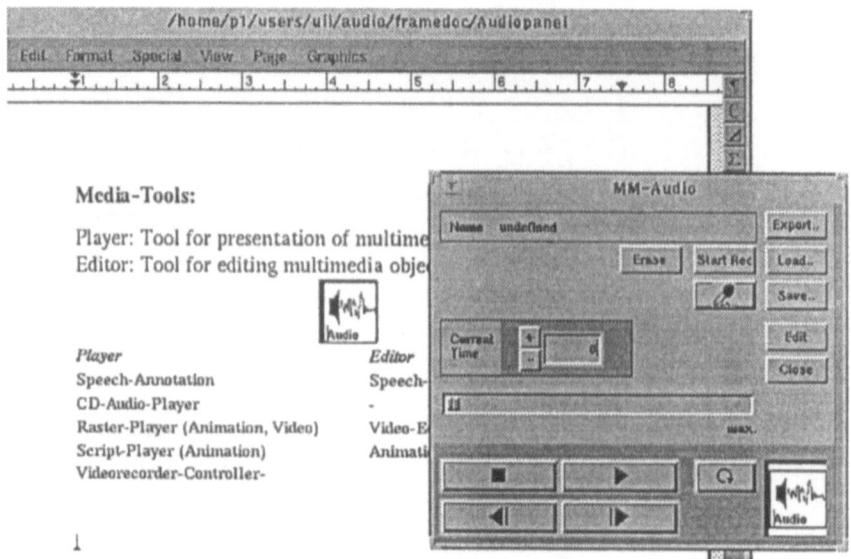

Figure 3. Voice Annotation Editor for the FrameMaker

The first prototype of the voice annotation editor was developed on a SUN SPARCstation2. The editor provides the functionality of an audio cassette recorder. Voice or sound can be recorded and replayed. The recording and replaying of sound is done using the audio device driver of the SUN SPARCstation2.

Currently, an improved version of the voice annotation editor is under development on Siemens-Nixdorf RW 320 (SiliconGraphics Iris Indigo) for FrameMaster V3.0, which is a Siemens-Nixdorf derivate of FrameMaker. The most important improvements are different levels of sound quality, recording of annotations with unrestricted duration, and a more sophisticated user interface. The implementation uses the audio device library of the operating system Irix 4.1.

A possible application szenario is as follows: Someone writes a FrameMaker document and would like a review of this document by a colleague. Therefore, he sends the document to the colleague via electronic mail. The colleague reads the document and adds his comments using voice annotations. For example, he could say "You should cite the paper of Miller here." or "This paragraph sounds a bit confused to me. Perhaps you could explain this point in more detail.", etc. Having added his comments the colleague sends the annotated document back to the author. The author can then listen to the voice annotations and improve his document according to the comments given by his colleague.

A similar application could be used in computer based training. A student who has a question about some part of the instructional material can put his question into a voice annotation and send the concerned page to the tutor. The tutor can answer the question using another voice annotation and send the annotated page back to the learner.

3 Speech Command Recognition

Using voice annotation the computer is only used as a medium to store, transport, and present spoken information. In order to improve human-computer interaction, however, the computer has to *understand* spoken information. Speech command recognition has to be used to recognize the spoken commands which have to be translated into the commands that can be interpreted by the application (see Figure 4).

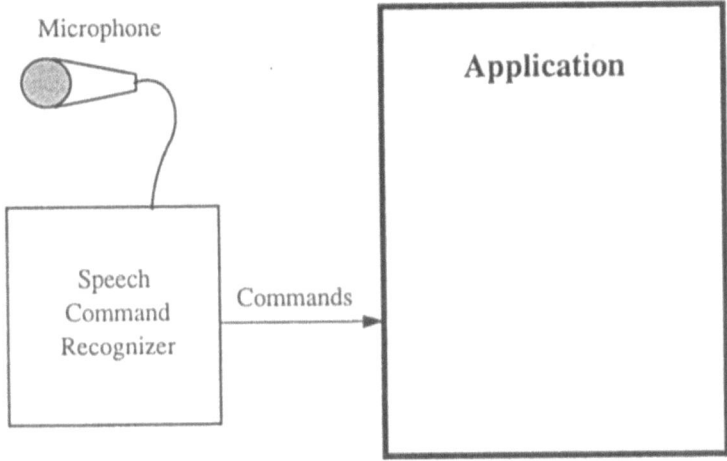

Figure 4. Speech Command Recognition

At the Siemens Corporate R&D Labs speech command recognition was tested in an animation system with a multi-modal dialog interface [NiHa91][NAH91]. The system realizes a simple virtual 3D world in which several objects can be created by reading object descriptions from a database. Besides the usual geometrical structure described by vertices and polygons, the objects also have physical properties which are represented by mass-points and constraints. Several attributes like elasticity, initial velocities, gravitation, and kind of presentation can be assigned to these objects and can be influenced by the animator during the session. Motion of the objects means that their movement is evaluated at each time step by an algorithm taking into account several constraints of the objects. The algorithm is based on the physical laws and approximates the motion dynamics by iteratively satisfying the constraints [NiLe90]. However, it considers only as many parameters as necessary, in order to keep the workload down. The simulation of the motion dynamics is performed in real-time and leads to motions of the objects which are accepted as natural by observers.

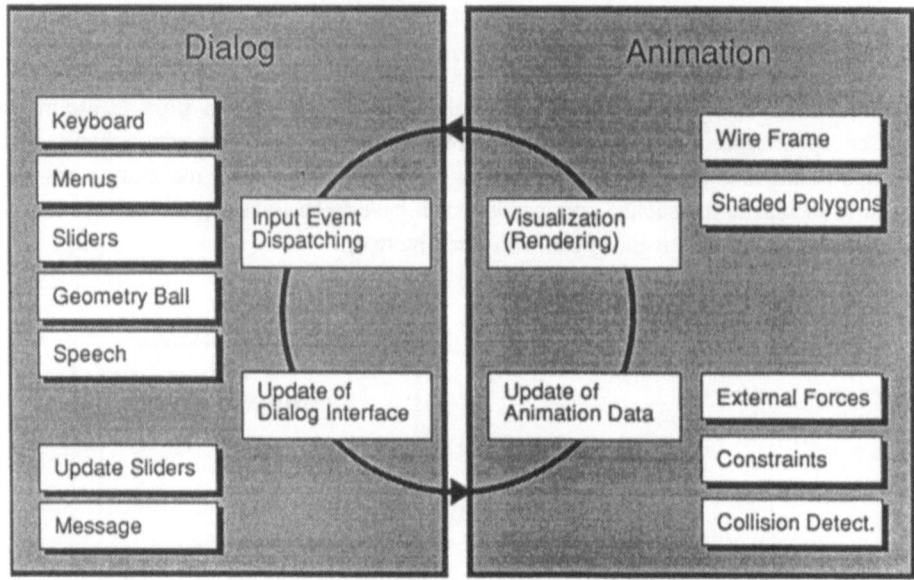

Figure 5. Structure of an animation system with
multi-modal dialog interface

The functionality of the animation system is separated from the dialog control as far as possible (see Figure 5). It is the task of the dialog control component to update the dialog interface, i.e. to give feedback to the user's inputs by showing the currently valid commands and the last chosen command, and to display error messages. An important feature of the system is that it offers several different input devices for interaction. There is conventional keyboard input, menus from which commands can be selected using the mouse, sliders for defining a position in 3D space, a geometry ball which is a 6D input device, and the speech recognizer. Independent of the input device by which a command was initiated, the dialog interface always has the same presentation and behavior.

Because several different input devices are offered concurrently, the user can select the input device which is most appropriate for the task at hand. Especially, the combination of two (or more) input devices promises good performance. The combination of speaking and pointing, e.g., is a very natural way of communication. Using speech input allows the user to keep his eyes on the object of interest while entering a spoken command. If the user wants to fix a vertex of a tetraeder in 3D space, e.g., he can stay with his eye focus and with the mouse cursor near the tetraeder while entering the spoken command "fix". Then he fixes the vertex by selecting it with the mouse.

For speech recognition the system uses a speaker-independent word recognizer using continuous Hidden-Markov-Models based on phonemes. This means that each word is recognized as a sequence of phonemes according to a pronunciation lexicon. The recognizer decides for the most probable phoneme sequence provided that some rejection criterion (probability too low) is not met. The speech recognition is independent of the speaker, i.e. the system does not have to be adapted to individual speakers. The recognition rate is about 98 % depending slightly on the actual command set.

With a special hardware it is possible to run the recognition process in real-time. This hardware consists of an acoustic front-end called AkuFE [AkHö89]. The AkuFE is a multi digital signal processor sytem which was developed within the speaker-adaptive continuous speech understanding system SPICOS [Zün90].

Today, we have reached a point where single word recognizers with small vocabularies (i.e. less than 100 words) are accurate and fast enough for commercial products. Technology is evolving very fast so that in the near future we will have recognizers for large vocabularies (i.e. more than 100000 words) and even for continuous speech in real-time [Zün90].

Conclusion

The integration of speech into workstation-based applications is now possible and useful, because it can improve human-human communication via exchange of electronic documents as well as human-computer interaction.

The important thing is that speech should not replace other communication media but should be a supplement. The right media mix is the key to success. Voice annotation, e.g., is an add-on which can improve the usability of many software products. However, it can be very time-consuming to listen to long voice annotations while a textual annotation could be glanced through very quickly.

Speech command recognition is very helpful if a dirty working environment does not allow to use conventional input media. Where conventional input media can be used speech command recognition can speed up the interaction because it reduces typing effort or cursor movements. Very promising is the combination of speaking and pointing as mentioned earlier.

Recognition of fluently spoken language requires high computational power. With the availability of more computational power the performance of speech recognition will improve in the near future. Speech recognition is useful for electronic query systems, e.g. itinerary information systems or hotel reservation systems, but also for applications like the "hearing typewriter" or automatic translation systems which allow people in different countries to communicate via phone using their individual mother tongue.

144

References

[AkHö89]   A. Aktas, H. Höge: Real-Time Recognition of Subword Units on a Hybrid Multi-DSP/ASIC Based Acoustic Front-End, Proc. IEEE International Conference on Acoustics, Speech and Signal Processing '89, Glasgow (1989), pp. 101-103.

[FM90]   FrameMaker V2.1, *Using FrameMaker*, Manual. Siemens Nixdorf Informationssysteme, User Documentation Department AP, Otto-Hahn-Ring 6, 8000 München 83, Germany

[FM90.1]   FrameMaker V2.1, *Integrating Applications with FrameMaker*, Manual. Siemens Nixdorf Informationssysteme, User Documentation Department AP, Otto-Hahn-Ring 6, 8000 München 83, Germany

[NAH91]   M. Niemöller, A. Aktas, U. Harke, U. Leiner, K. Zünkler: SESAM: A Prototype Multimedia System combining Computer Animation with a Speech Dialog, Proceedings of the 4th Interantional Conference on Human-Computer Interaction, Stuttgart (1991)

[NiHa91]   M. Niemöller, U. Harke: Combining Computer Animation and Speech Understanding to a Multimedia System, Proceedings of the International Conference on Communications ICC'91, Denver, Colorado, June 23-26 (1991)

[NiLe90]   M. Niemöller, U. Leiner: Approximative, Dynamic Simulation of Objects in Computer Animation Systems, Proceedings of Computer Graphics '90 Conference, London, Nov. 6-8 (1990)

[Zün90]   K. Zünkler: Speech-Understanding Systems: The Communication Technology of Tomorrow, in Schwärtzel and Mizin (eds.): Advanced Information Processing, Springer-Verlag, Berlin (1990), pp. 227-251.

Multimedia: A New Eye-Catcher In An Evolutionary Process

Multimedia: a new focus in an evolutionary process

J. Samuel

Siemens Nixdorf Informationssysteme AG

1. Introduction

Ladies and gentlemen,

Thank you for allowing me to speak to you on the subject of multimedia.

Multimedia   is a method, by which more effective communication is achieved through the computer-aided integration of audiovisual media such as text, (animated) graphics, speech as well as still and moving pictures   acc. ISO: the property of handling seseral types of representation media

According to the ISO definition, a multimedia presentation is characterized by handling several types of representation media.

For me, this definition is unsatisfactory. Therefore, I would prefer to work with the following definition:

Multimedia is a method by which more effective communication is achieved through the computer-aided integration of audiovisual media such as text, (animated) graphics, speech, as well as still and moving pictures.

So I would prefer to put the aim of a better communication in the forthground!

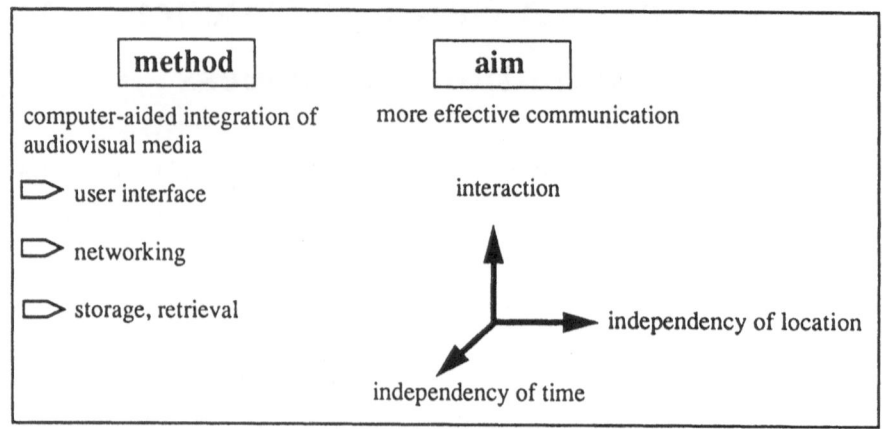

Communication between human beings is always most effective if those participating in the communication process can interact through seeing, hearing and doing.

Today, we have come together in the same place at the same time in the context of this symposium on "Intelligent Workstations". I have taken on the task of informing you on the subject of multimedia as one aspect of intelligent workstations.

Starting from the knowledge that information is most effectively transported by using a combination of sight, sound and action, my procedure is not to

- show you transparencies without explaining them,
- nor to simply speak without transparencies.

Rather, I use both speech and transparencies as my media. In order to include the aspect of action as well, I invite you to ask questions so that we can enter into genuine interaction.

Communication can take place not only when

- people are in the same place at the same time,
- but also when they are in different places at the same time
- or at the same place at different times
- or at different places at different times.

In the last three cases, communication is especially effective if it includes sight, sound and action, thus achieving a high degree of interaction. If this interaction is to be computer-aided at every point of the space/time frame, it involves the following aspects of information technology:

- User interfaces,
- networks and
- data storage.

```
Agenda

  ▷  introduction

  ▷  technological aspects

  ▷  market aspects

  ▷  SNI's offer
```

Consequently, I would like to devote the second part of my talk to describing the technological aspect of multimedia. In the third part, I will focus on the multimedia market. In the fourth part, I will explain how Siemens Nixdorf is operating in the multimedia market.

As the product planner in charge of user interfaces and multimedia at Siemens Nixdorf, please do not expect me to give you a detailed lecture on technology or marketing research, but rather a general overview on the subject of multimedia.

2. Technological aspect of multimedia

As I see it, the computer-aided integration of audiovisual media in the areas of user interfaces, networks and data storage/retrival involves four basic requirements:

```
basic requirements:

  ▷ The different media must be processed digitally
    (transition period: hybrid system conbining digital and analog technology)

  ▷ User interfaces must become multimedia-capable

  ▷ It must be posible to transfer digitized multimedia data in real time

  ▷ The storage of digitized multimedia data must be cost-effective
```

1. The different media must be processed digitally.

This does not exclude a transition period in which hybrid systems combining digital and analog technology are used, since f.e. a higher video quality is still being achieved at the present time through analog technology.

2. User interfaces must become multimedia-capable.

I consider this to be an evolutionary process. Let's review the development of user interfaces:

The first user interfaces were alphanumeric and operated exclusively in block mode. With these user interfaces, a special transfer instruction served to pass a fully edited screen to the computer for processing. The computer then presented the result of processing on the screen. The alphanumeric user interfaces in block mode were eventually expanded or replaced by alphanumeric user interfaces with single-character processing. This already served to increase interaction between users and computers. A further step toward increased interaction was the introduction of graphical user interfaces, which substantially broadened the user's scope of action. Multimedia user interfaces will enrich the communication between users and computers, particularly on an emotional level - consider music or moving pictures - thus increasing interaction even further. In this respect, multimedia is nothing new, but is simply a new **eye-catcher evolutionary process.**

3. It must be possible to transfer digitized multimedia data in real time.

This requirement is based on the fact that highly interactive communication must also be possible between users who are at different locations.

4. The storage of digitized multimedia data must be cost-effective.

This requirement arises from the desire to buffer a multimedia message and be able to call it up at some time in the future.

I would like to discuss these last two requirements in detail:

The problems encountered in multimedia user interfaces involve data transfer rates and datasets.

For example, one second of video at a resolution of 512 x 480 pixels, a 3-byte color depth and 30 frames per second results in an enormous dataset of 22 Mbytes; an hour of such video then requires 80 Gbytes.

Even within the local architecture of desktop systems, the data rates are far below this level. About 4 Mbytes per second can be transferred on a PC bus; data is fetched from a hard disk at a rate of 5 Mbytes per second; even from a CD-ROM, it can be fetched at a rate of no more than 150 kbytes per second. In local area networks such as Ethernet, rates of approximately 10 Mbits per second are possible, the token ring can reach speeds of 16 Mbits per second and FDDI is specified at 100 Mbits per second.

The storage capacity of a well-configured PC hard disk of, for example, 330 Mbytes or even a 650-Mbyte CD-ROM can store no more than a few seconds of video.

The problem in multi-media user interface	user-interfaces		The solution to this problem		user-interfaces	
	1 sec video: 512x480x3Bx30 = 22 MB 1 h video: = 80 GB		is data compression video :150 still image :10-15	:1 :1	1 sec video: 22 MB : 150 = 150 KB 1 h Video: = 540 MB	
involve data transfer	transmission				transmission	

	band-width:
	PC- BUS 4 MB/sec
	     Harddisk 5 MB/sec
	     CD-ROM 150 KB/sec
	Ethernet 10 Mb/sec
	Token Ring 16 Mb/sec
	FDDI 100 Mb/sec
					band-width:
					
					CD-ROM
					Ethernet 1 Video Stream
					Token Ring 5 Video Stream
					FDDI 10 Video Streams
					                 55 Video Streams	

and data sets	Storage				Storage	

	capacity:
	
	PC-Harddisk
	CD-Rom 330 MB
	                 650 MB
					capacity:
					
					PC-Harddisk
					CD-Rom 37 min.
					                 72 min.	

The solution to this problem is data compression. For a moving picture, this means no longer transferring or storing a video sequence for each image, but only taking into account changes with relation to the previous image. In this way, for example, video data can be compressed by a factor of 150.

The dataset for 1 second of video is then reduced to a mere 150 kbytes; an hour of video is 540 Mbytes.

These dimensions are adequate when reading a video stream from a CD-ROM. If you also include the corresponding loading in the calculation, it would be possible to transfer about 5 video streams simultaneously via an Ethernet, about 10 via a token ring and about 55 via FDDI.

Approximately 37 minutes of compressed video can be stored on a well-configured 330-Mbyte PC hard disk and around 72 minutes on a 650-Mbyte CD-ROM.

Special algorithms are applied when data is compressed or decompressed. Some of these algorithms are not standardized, such as INTEL's RTV (real-time video) or the qualitatively superior PLV (production-level video); others are in the process of standardization. The most important of these are JPEG (joint photographic expert group) for stills and MPEG (moving pictures expert group) for moving pictures. Regardless of the algorithm used, there is an important trend toward chip-supported compression and decompression of video data using programmable video processors. A single PC with a programmable video processor will be capable of processing several algorithms.

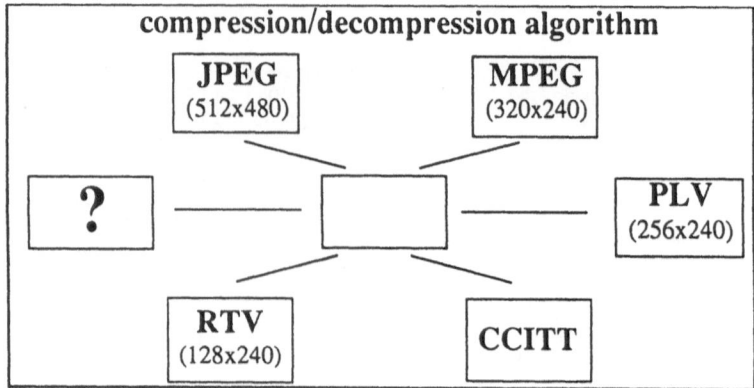

3. The multimedia market

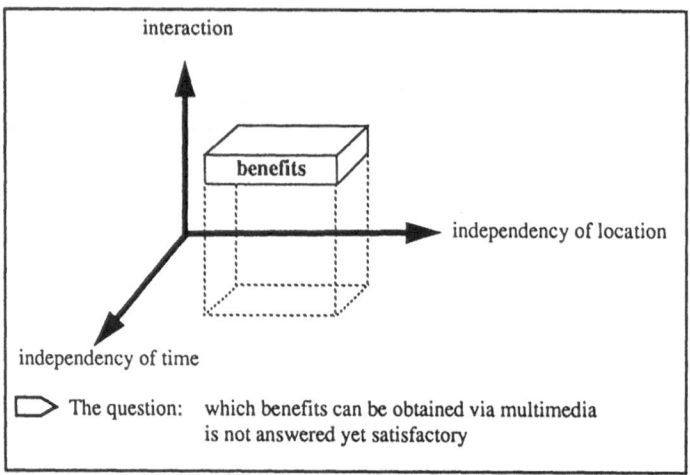

I begin on a sobering note.

The question of the purposes that multimedia is intended to serve is not answered satisfactory yet. Certainly, there are a number of possible market segments such as training, education, publishing, presentation, self-service, communication, home & consumer and entertainment, but no well-founded statements have been made as to the possible size of these segments. In particular, forecasts by various marketing research institutes as to the overall volume of the multimedia market differ greatly and are somewhat indefinite. The Dataquest figures for 12/91 seem relatively realistic, predicting a volume of 8.5 billion dollars for the overall multimedia market in 1992. According to Data-quest, this volume will expand to 12.5 billion dollars by 1996.

- Philips builds its multimedia program around the CD. It is developing the CD-I (compact disk interactive) in cooperation with Sony.
- Intel is developing DVI (digital video interactive) in order to provide PCs with full-motion video capability.
- Microsoft is attempting to set multimedia standards and introduce new software packages on the market.
- Finally, Apple is expanding its graphically-intensive Macintosh computer to process video and hi-fi sound.

4. SNI in the multimedia market

> **SNI intends to participate in the multimedia market but we will enter it cautionsly and with a view to proper timing**
>
> ▷ Multimedia hardware platforms
>
> ▷ Multimedia tools
>
> ▷ Multimedia application software
>
> ▷ Multimedia solutions

After reading the signals being given off by the multimedia market, we fully intend to participate in this market. However, we will enter it cautiously and with a view to proper timing. Our range of products, when fully developed, will include the following areas:

1. Multimedia hardware platforms
2. Multimedia tools
3. Multimedia application software
4. Multimedia solutions

Before closing, I would like to deal with these four areas in detail.

Re 1. Multimedia hardware platforms

We will give first priority to PC-based multimedia hardware platforms. These are:

- The multimedia PC (MPC), which we will make available next month. This is a standard PC equipped with a CD-ROM and an audio board; a microphone and loudspeaker will be op-

152

> The multimedia market has not yet been adequately analyzed
>
> POI, POS, CBT appear to be key segments for multimedia
>
> Multimedia is a growth market

For this reason, at this point we can make only the following three statements about the multimedia market:

1. The multimedia market has not yet been adequately analyzed, as is directly evident from what I've already said.

2. The point of information (POI)/point of sales (POS) and computer-based training (CBT) market segments appear to be the key segments for multimedia. Despite the indefiniteness of the market figures, the different marketing research institutes have reached a qualitative consensus on this point.

3. Multimedia is a growth market. Its growth rate is again perceived very differently; therefore, I will limit myself to these qualitative statements

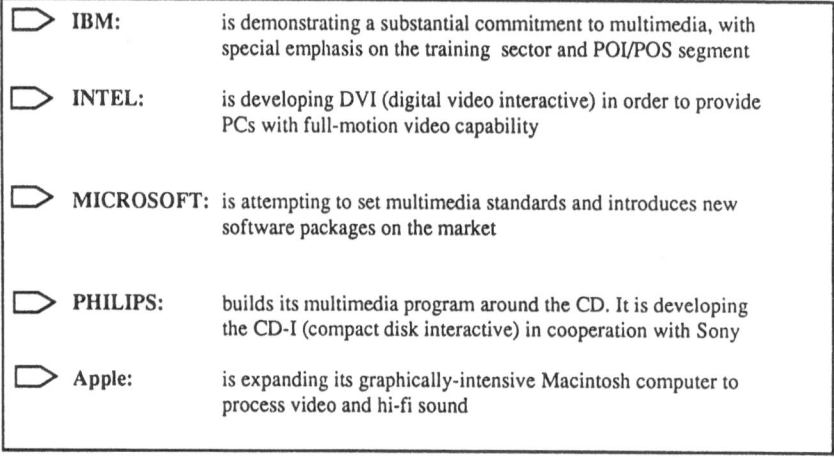

> IBM: is demonstrating a substantial commitment to multimedia, with special emphasis on the training sector and POI/POS segment
>
> INTEL: is developing DVI (digital video interactive) in order to provide PCs with full-motion video capability
>
> MICROSOFT: is attempting to set multimedia standards and introduces new software packages on the market
>
> PHILIPS: builds its multimedia program around the CD. It is developing the CD-I (compact disk interactive) in cooperation with Sony
>
> Apple: is expanding its graphically-intensive Macintosh computer to process video and hi-fi sound

Our competitors in the multimedia market follow various strategies compatible their own traditional markets:

- IBM is demonstrating a substantial commitment to multimedia, with special emphasis on the training sector and POI/POS segment.

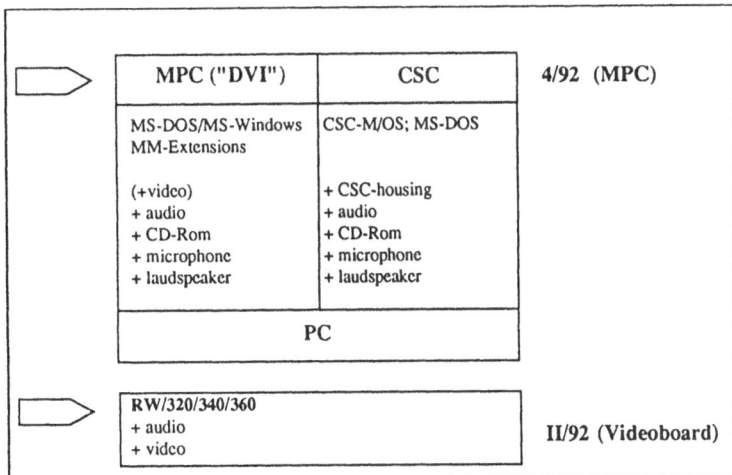

tionally available. The operating system is MS Windows with Multimedia Extensions. The

second step will be to expand our MPCs to include video capabilities.

- Our CSC (Custer Self Service Center) devices are already available with multimedia features. However, we plan to increase the affinity between the MPC and CSC by using the same boards in both devices (e.g. audio board).

Our second priority will be our RISC workstation family, comprising the RW320, 340 and 360, 460 which are upgraded to include audio capabilities. In the second quarter of this year, a video board will be available for this workstation family.

Re 2. Multimedia tools

With CADT, we have long had an authoring tool in the CSC range. However, we now operate on the assumption that the trend is moving toward authoring tools for the standard PC or multimedia PC. We also intend to take advantage of the affinity between CSC and PCs with regard to multimedia tools. After careful evaluation, we will introduce a strategic authoring tool for CSC and MPC. To bridge the time until we actually introduce this tool, we will introduce the author system Authorware as an imediate measure.

```
┌─────────────────────────────────────────────────────────────┐
│  │ situation │                                               │
│                                                               │
│   -  CSC: CADT                                                │
│   -  PC: From here comes the momentum                         │
│  │ aim │                                                      │
│                                                               │
│   -  Take advantage of the affinity between CSC and PC's      │
│      One strategic authoring tool for CSC and  PC's           │
│  │ strategy │                                                 │
│                                                               │
│   -  Careful evaluation                              IV/92    │
│   -  introduction of strategic authoring tool        I/93     │
│   -  introduction of Authorware as immediate measure II/92    │
│                                                               │
└─────────────────────────────────────────────────────────────┘
```

Re 3. Multimedia application software

```
┌─────────────────────────────────────────────────────────────┐
│  enrich suitable application software with multimedia features │
│                                                               │
│   ⊏▷ Multimedia components will profit, by using application software │
│      as ideal carier systems                                  │
│                                                               │
│   ⊏▷ The application software will profit, by being rendered more marke- │
│      table trough multimedia components                       │
│                                                               │
│   ⊏▷ Concentrating on multimedia features in which the complexity of │
│      hardware and system-software is easily mastered          │
│                                                               │
│   ⊏▷ Providing multimedia system-software that is aligned to the │
│      application software                                     │
│                                                               │
│   ⊏▷ View on cost-effectiveness                               │
└─────────────────────────────────────────────────────────────┘
```

We will enrich suitable application software with multimedia features. The application software and multimedia components must both profit from this enrichment. The multimedia components will profit by using application software as ideal carrier systems; the application
software will profit by being rendered more marketable through multimedia components. Since
our application software is particularly aimed at client-server architectures, we will begin by
concentrating on multimedia features in which the complexity of the hardware and system-
software infrastructure is easily mastered. We will provide multimedia system software that is
aligned to the application software. Every investment in the development of multimedia components for application software will be made with a view to cost-effectiveness.

Re 4. Multimedia solutions

Examples

▷ Präsentation of the product range using multimedia

▷ CBT-programm about multimedia using multimedia

Our company has already developed a number of multimedia approaches. One example is our Business System Division's presentation of our product range using multimedia; another is a training approach dealing with the subject of multimedia. In both cases, we also use DVI technology. In the future, we plan on increasing the number of such solutions and using them more intensively within our own company.

We will also seek more opportunities for cooperation in order to make our full product range available as quickly as possible.

Architectural Trends in Workstation Design

Michael Geiger
Siemens AG

Abstract

The conception and development of intelligent systems for professionals is doubtlessly one of the biggest challenges of the century for the computer industry. Today most approaches deal with new information processing and programming techniques based on general purpose computing platforms. Future intelligent systems will either use completely new system architectures, like for example neuro-computers or systems based on fuzzy logic, or will use traditional systems equipped with a variety of new functionality, e.g. multimedia technology.

Among all computer architectures available today, the workstations offer the best potential to be efficiently used by intelligent systems. In comparison with Personal Computers (PCs) for example they are superior in terms of processing power, graphics functionality, networking, peripherals and operating system; compared to mainframes they offer better usability, flexibility and scalability.

Starting with a list of general requirements for future workstation architectures, the paper deduces corresponding design trends. It focuses on hardware and system software trends; aspects specific for application software (e. g. user interface management systems) are not within the scope of the paper. Everything discussed in the paper is derived from a technical point of view, not from a marketing one. Finally it must be taken into account that there is no common time frame for the presented trends. Some trends are more or less reality, i. e. first implementations are already available. Other trends impose grave technological problems so that their realization will not take place within the next 2 - 5 years. And there are some trends which will not influence workstation generations before the next century.

1 Future Workstation Requirements

When asking system programmers, application programmers or users about their requirements for future workstation generations, the same set of basic requirements is mentioned: better performance, enhanced functionality and standardized computing platforms. After a closer look into theses issues a variety of different areas appear. Higher performance, for example, is not exclusively limited to general purpose computing power, but also involves graphics, networking or signal processing performance. This is also true for standardization issues where besides the hardware environment, also the operating system, the application programmer's interface or the underlying window system are involved.

In contrast to the more conservative and well known requirements mentioned above the longing for enhanced functionality opens new horizons in the workstation world. One of the decisive tasks of the decade is the seamless integration of multimedia functionality into existing workstation architectures. Although first steps have already been undertaken a lot of powerful components still have to be designed and implemented (e.g. enabling components for sophisticated multimedia applications like computer supported cooperative work, CSCW).

Since it is impossible to cover all facets of the trends resulting from the requirements mentioned above, the paper concentrates on seven global areas shown in figure 1.

2 Trends in Workstation Design

As a matter of fact there is no bunch of isolated trends, but a closely coupled network of architectural dependencies, where a move in one direction strongly influences other issues. A good example is multimedia. If, for example, a camera has to be attached to the workstation to record a speaker in a video conference, it is inevitable to have at least a frame grabber available. Due to the requirements of the network application, the video subsystem should allow for digital processing of the video data. The vast amount of video data cannot be transmitted over a computer network without compression. When using advanced compression techniques appropriate hardware must be available since today realtime compression fulfilling high quality requirements cannot be performed in software only.

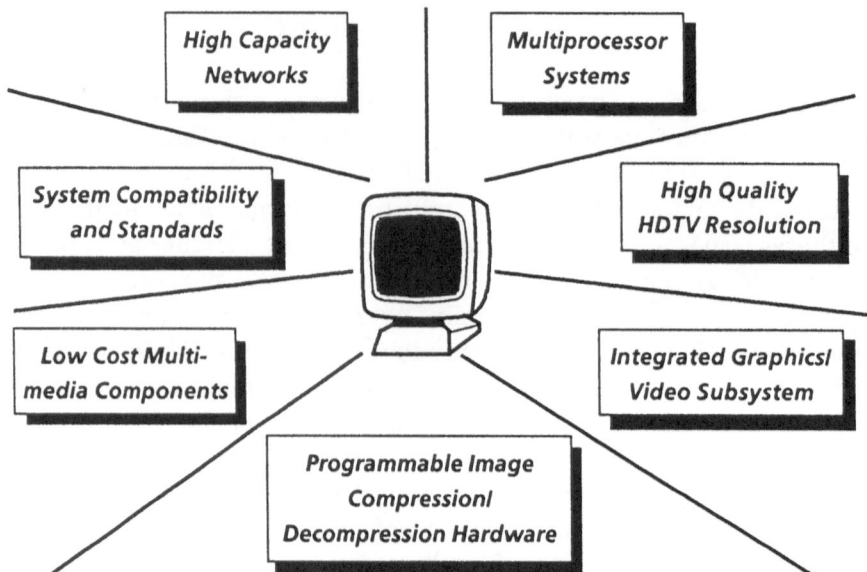

Figure 1: Trend Setting Areas for Future Workstation Architectures

Furthermore, standards are inevitable, concerning as well compression schemes as the programming interface for the multimedia application. The list could be continued for some time.

Consequently, the following chapters describe trends for the design and implementation of **basic** workstation features of the future. Powerful systems must be based on a balanced concept including all of them.

2.1Integrated Graphics/Video Subsystem

Since the migration from alphanumerical output to graphics interfaces in the 1980ies, the graphics subsystem has evolved into a central component of today's workstations. During the last years several workstation manufactures have added 3D functionality to their traditionally 2D-oriented graphics systems. Within the coming years hardware supported 3D graphics subsystems will be an integral part of all workstations, wether low-end or high-end system.

In order to provide powerful 3D functionality it is necessary to integrate a hardware supported graphics pipeline (rendering pipeline) into the workstation. Thus the necessary floating point performance is granted without loading the host CPU. Additionally, the pixels are generated close to the frame buffer so that the copying of pixel data between the system memory and the graphics subsystem can be avoided. As a result, the I/O-bus is not burdened. With hardware supported 3D graphics subsystems the rendering of complex animation scenes in real time and the usage of computationally intensive rendering techniques like ray tracing become feasible. Typical application areas are mechanical CAD, multimedia presentations and virtual reality.

Apart from 3D graphics, video is at the threshold to evolve into a standard feature of modern workstations. Starting with displaying monochrome, grayscale or color images on the workstation monitor, more and more of today's state of the art workstations are capable of capturing and displaying motion video sequences and the associated audio. Figure 2 shows three basic implementation methods.

- **Analog Overlay**
  Here graphics and video data are processed completely independent of each other. Analog video information is captured from the video camera and then overlayed with analog RGB signals from the graphics hardware before being displayed. The major advantage of this approach lies in its simplicity. In return there are several disadvantages: only fix-sized windows can be supported and the editing of video data is impossible since the data is not available in digital form. The analog overlay solution is not suitable for workstations, but is realized by various PC cards because it is easy to implement.

- **Decoupled Architecture**
  With this solution the captured video data is digitized and stored in the video card's frame buffer. As a result, signal processing algorithms can be applied to the video data, and variable-sized windows can be supported. Unfortunately, the resulting pixel data must be copied over the I/O-bus to the graphics subsystem, thus creating a severe system bottleneck. Since the decoupled solution is quite easy to implement and rather flexible at the same time, it is realized with several workstation video cards (namely Sun Microsystem's VideoPix).

- **Coupled Architecture**
  The coupled architecture offers the most powerful and sophisticated solution of an integrated graphics/video subsystem. Here, digitized video data and graphics data are stored in the same frame buffer. As a result complex algorithms can be applied to the merged data realizing

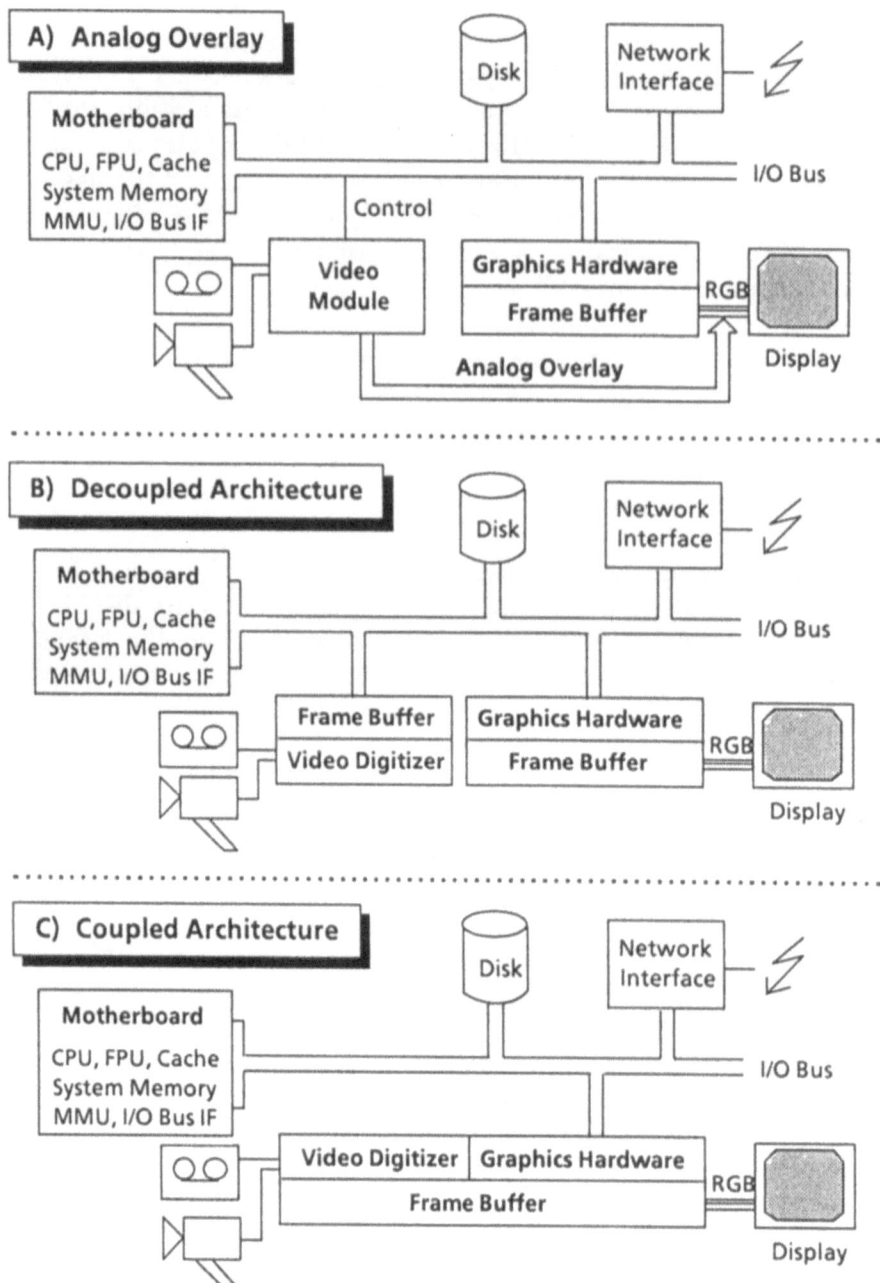

Figure 2: Basic Implementation Methods for Motion Video

enhanced features like overlayed graphics and video in one window, transparency effects, or texture mapping with live video on 3D objects. Other advantages comprise cost and space reduction for the frame buffer as well as avoiding the bottleneck I/O-bus. The main reason why this solution (which is definitely best suited for workstations) is not yet realized are grave implementation problems like the bandwidth bottleneck on the frame buffer or synchronization issues. Additionally, such a configuration misses flexibility since graphics and video subsystem must fit together, i.e. an upgrading of an existing workstation platform (including the graphics subsystem) with a video subsystem is very difficult.

Future solutions will not only integrate graphics and video on a dedicated subsystem to be plugged into the system bus, but will be directly integrated on the motherboard. Thus the system bus is no longer involved, and the realization of portable workstations with multimedia capabilities is feasible.

There are numerous potential application scenarios for such systems, including for example the large area of office applications and multimedia presentations.

2.2 Programmable Image Compression/Decompression Hardware

The integration of video functionality into the workstation is one thing, the efficient handling of the video data and the exchange of real time video between different sites is something else. The vast occuring data rates make compression a necessity, and since software compression is not feasible with high quality real time video, the use of dedicated compression/decompression hardware is inevitable. Moreover, codec programmability is required since there are a lot of (more or less) standardized compression schemes (e.g. JPEG, MPEG, H.261, etc.) which must be supported, and new ones are still forthcoming (e.g. fractal compression).

The next step towards applications like Computer Supported Cooperative Work (CSCW) requires the parallel processing of multiple video streams using different compression algorithms. Thus, dedicated multiprocessor hardware able to handle multiple video streams simultaneously is needed. Additionally, a multiprocessor solution can offer better image quality in terms of higher resolution or a higher frequency of frames per second.

Finally, the codec functionality will be integrated on the host CPU as on-chip coprocessor, thus guaranteeing better performance than specialized codec subsystems. An implementation of the graphics/video subsystem could then use the system's memory bus instead of the I/O-bus. Along with an integrated graphics/video subsystem on the motherboard, cost and space reduction can be achieved. An example might be an i80x86 with DVI functionality on it. But although the concept is very promising products will not be available within the next 5-7 years.

2.3 High Quality HDTV Resolution

Today's video resolution ranges from stamp-sized 8bit coded color or grayscale windows to CIF (352x240, 16bit color) or VGA (640x480, 16bit color) resolution. High end graphics/video subsystems of tomorrow will support High Definition TeleVision (HDTV) resolution (e.g. 1920x1152, 24bit color). Due to the vast bandwidth needed on the I/O-bus (about 1.2 Gbit per second, which is approximately a factor of 8 compared to VGA resolution), appropriate hardware supported compression/decompression is mandatory. Another bottleneck is the frame buffer, where a pixel has to be read or written within 20 ns. In principle, all applications requiring large high resolution displays (e.g. ECAD, MCAD, medicine, multimedia presentations) are targeted.

When HDTV will be established in the consumer electronics world as the new broadcasting standard (which is rather likely to come true within the next years), and the needed codec hardware for the computer environment will be available, both worlds can be combined to a new system class, the "tele-workstation". Such a new system concept would definitely require a digital HDTV norm in the broadcasting area as it is proposed in the USA. Typical applications are distant learning or "video on demand".

2.4 Low Cost Multimedia Components

Although multimedia features are widely spread in state of the art work-stations there is a definite lack of low cost multimedia components. New applications like multimedia mail or CSCW can only be used effectively if all potential partners are able to process the new data types (namely motion video and audio). Therefore future workstations must be equipped per default with efficient (and cheap) video subsystems, hifi audio modules and high capacity optical storage media like CD-ROM or CD-MO. Additionally these subsystems must be standardized to allow applications to be used in heterogeneous systems. Typical application areas are hypermedia applications and multimedia databases.

An additional requirement is the ergonomical integration of these components into the workstation. Future systems have to provide displays incorporating basic multimedia modules like video camera, speakers and microphone. Novel input devices like the DataGlove for three-dimensional picking or the VideoGlove must be supported. Traditional displays will be replaced by touchscreens and flat panels. Keyboard and mouse operation will be complemented by pen input. Applications can be found above all in the office environment.

2.5 High Capacity Networks

With the implementation of netwide applications new high capacity optical networks are required, which guarantee an appropriate sustained bandwidth. Today FDDI is the state of the art solution for in-house networks. With the realization of workstation adapter cards for public broadband networks (e. g. ATM subsystems) completely new ways of computing are feasible. In the long run ATM networks will be used as in-house networks, and thus maybe make Ethernet obsolete. The classic bottleneck of today is most often the network. With broadband optical networks the bottleneck will be shifted towards the protocol stack. Therefore new high-speed protocol stacks are needed, which grant effective operation regarding the transmitted data (e. g. no retransmissions with video data). Optical networks are for example required by CSCW applications or when realizing worldwide video servers.

The next step is the usage of the ATM network as workstation bus. The result is a "distributed workstation" (rather: a distributed system) where functional units are connected via the network instead of assembling all modules in one cabinet. Such a system concept allows net-wide resource sharing. A central requirement for such a realization is a distributed operating system which guarantees efficient and transparent resource management. Such a distributed system enables the implementation of global information systems spread all over the world.

Another step towards modern communication means is mobile telecommunication. Then the realization of networked portable workstations is feasible. Although there are a lot of basic problems to tackle (e. g. bandwidth bottlenecks or health perils inflicted by electro-magnetic fields), the needed technology will be available within the next 10 years.

2.6 Multiprocessor Workstations

In the past there was a clear distinction between (monoprocessor) workstations and highly parallel systems (using for example the hypercube architecture). Today a new class of multiprocessor workstation begins to establish itself as general purpose systems in between, trying to supply as much performance as dedicated parallel machines. Unfortunately there are a lot of basic problems to be solved before this target can be reached. Typical multiprocessor systems offer 4 - 8 processors. The next step is the integration of more than 16 CPUs into the workstation. Such an integration creates several bottlenecks, especially on the memory bus. Sophisticated cache coherency protocols must be realized. Today most multiprocessor workstations use a symmetric approach. Future systems might be realized in an asymmetric way. Another unsolved question is the use of distributed memory versus shared memory. But even if all bottlenecks and problems arising with the integration of 16 CPUs into a single workstation can be solved efficiently, there is a lack of parallelizing compilers which produce adequate object code for such an architecture.

In a further phase these multiple processing units will be integrated on a single chip to design a "multiprocessor on a chip".

Although all the problems mentioned above are already considered by workstation designers, and first solutions are not too far ahead, there are several basic issues which endanger a global success of multiprocessor systems in the workstation market:

1) *Can multiprocessor workstations compete with comparable system architectures?*
   - there is very little (appropriate) SW available
   - multiprocessor workstations are very expensive (bad cost/ performance ratio)
   - monoprocessor performance is increasing steadily
   - considering performance issues, highly parallel machines are still far ahead

2) *Will appropriate processors be available in the future?*
   - high speed communication interfaces between processors must be implemented
   - the processor's instruction set must offer support for cache coherency protocols

3) *Will appropriate bus systems be available in the future?*

- high performance bus concepts must be implemented
- busses must provide a very large bandwidth in order to avoid being system bottlenecks
- new standards are necessary since standardized bus systems of today cannot satisfy the requirements

Summarizing, multiprocessor workstation today define the high-end of the workstation segment, but it is still uncertain if they can establish themselves permanently in the workstation market of the future.

2.7 Overall System Compatibility and Standards

Throughout the preceding chapters a lot of dedicated (and often isolated) hardware issues have been discussed. In order to connect these modules and to allow efficient interworking, the operating system has to supply adequate functionality. Thus, an enhanced operating system guaranteeing real time capabilities and specific synchronization schemes is required. Novel input devices like pen input for example must be seamlessly integrated into the operating system. The best attempt to meet the requirements might be the use of lightweighted microkernel architectures (like for example the MACH kernel).

Another decisive trend is the support of worldwide standards. Above all, standardized Application Programmer's Interfaces (API's) must be available. When realizing networked applications, standardized networking conventions (like e. g. SUN Microsystem's RPC-protocol or the OSI stacks) are required. The development of shrink-wrapped software can be enforced by using binary standards like the Application Binary Interface (ABI) or the Architecture Neutral Distribution Format (ANDF). Finally, the support of generally acknowledged data exchange formats (e. g. the Office Document Architecture, ODA) is inevitable.

In the long run, there is one paramount goal: interoperability of all applications across heterogeneous systems (applications like Sun Microsystem's Network File System, NFS). When it does no longer matter wether an application is running on a PC, a workstation or a mainframe, the old dream of really open systems will be realized. I guess we all hope to be still in the business when it finally happens.

3 Conclusion

The paper has presented many trends different related to hardware and system software for future workstation design, but considering the whole impact workstations have on the computer industry it must be admitted that a lot of issues have been excluded from discussion. Summarizing the last chapters four general trends can be seen:

- Processor performance will be doubled every two years
- Multimedia functionality will be an integral part of future workstations
- Workstation design aims at smaller, portable systems
- The final goal is the melting of PC, workstation and mainframe architectures into truly open systems

Pinpointing the differences between various system architectures nowadays is often difficult. Tomorrow, a classification into different system architectures will be impossible. And the day after tomorrow, the traditional computer systems will be replaced by a pool of networked, worldwide resources, allocated on demand by intelligent agents, which present the appropriate information to the user on his/her personal workstation. The era of intelligent computing has just begun.

References

[1]   USENIX Association (ed.): "Multimedia - For Now and the Future", USENIX Summer Conference Proceedings 1991

[2]   D. F. Foley, et.al.: "Computer Graphics Principles and Practice", Second Edition, Reading, Massachusetts, Addison-Wesley 1990

[3]   K. Braun, M. Geiger: "Architectural Concepts for Multimedia Systems", Siemens Internal Report, ZFE ST SN 1-133, 1991

[4]   C. Müller-Schloer, E. Schmitter (ed.): "RISC-Workstation-Architekturen", Springer-Verlag Berlin 1991 (german)

[5]   H. Johnen, M. Östreicher, W. Woborschil: "HDTV und Datenverarbeitung", Siemens Internal Report, ZFE ST SN 1-141, 1992 (german)

Physicians' Use of Intelligent Systems

James C. Felli, MS and Albert H. Rubenstein, PhD
Northwestern University

Abstract

We consider the concept of intelligent machines and their present availability within the medical profession in the United States. Pursuant to considering features of intelligent systems needed over the next five to ten years, we look into barriers to their acceptance by the medical community. These barriers are technological and economical, usage-based, and cultural in nature.

Motivation

The explosion of the computer industry over the last two decades has brought with it several notable advancements in medical technology. Where, prior to the advent of personal and affordable computers, medical professionals were forced to rely upon sheaves of site-dedicated documentation and slow communication channels, there now exists the opportunity for near-instantaneous communication and the combined graphical representation of data extracted from a great many distinct sources. And yet, "Despite the major advances in the science and technology of health care, the training of physicians is little different today from what it was half a century ago [2]." There have, of course, been some changes involving the introduction of electronic support.

The Center for Information and Telecommunication Technology (CITT) at Northwestern University has undertaken a theory-based, empirical study involving samples of professionals called the UNIS (Users' Needs for Intelligent Systems) project [7]. The UNIS project was designed with three goals in mind [8]:

- to improve understanding of a professional's work including such things as the way in which they perform their tasks, problems they encounter, and their needs for new and improved intelligent support.

- to identify opportunities for new and enhanced electronic support for the professional in the form of systems, software, and machines that can increase the professional's productivity and effectiveness.

- to consider the new ideas in the context of intelligent support systems which are currently under development and explore possible find matches for the user's needs over the next five to ten years.

To date, the study has focused on a sample of professionals from these groups: architects, lawyers, physicians, scientists and engineers, and financial analysts. The considerations and concerns presented in this paper are the direct result of the field studies phase of UNIS. All propositions, conclusions, and inferences have been drawn from formal, in-depth interviews with nine physicians out of a total of 38 interviews from the five groups of professionals. Unless specifically cited to the contrary, the direct quotations and proposition statements included within this paper have been drawn from these interviews. In addition, over 150 propositions have been extracted from these interviews which indicate important attitudes of the medical community and other professionals toward the intelligent machines of today and tomorrow.

Introduction

In keeping with our goals of identifying new opportunities for intelligent machines, as well as ascertaining desired qualities of future systems, we did not make a priori assumptions regarding our interview subjects' basic concepts of intelligent systems.

Our first task was one of definition. Before we could properly gather information from which to determine the prevalent attitudes of medical professionals toward intelligent machines, it was necessary to come up with a working definition of such machines. The subsequent use of this definition provided a common basis for discussions with interviewees and allowed us to conduct our investigations in a more straightforward manner.

The UNIS project uses the following as its definition of intelligent systems:

> *"Intelligent systems are all computer based hardware and software which provide assistance to people in the conduct of all their professional activities requiring the exercise of intellectual skills* [7].*"*

As expected, some of the respondants voiced several objections to this proposed definition. Foremost was that of scope. Most of the physicians interviewed felt the definition to be too limited -- they contended that such systems should go beyond merely rendering information or mechanical aid and actually *extend the abilities of the user beyond normal levels*. While the UNIS definition inherently implies that a user's efficiency is enhanced, the argument made by doctors is that *intelligent* systems should somehow increase the user's operational level of expertise, allowing the user to function better and more effectively than could otherwise be expected. That is to say, users of such intelligent systems should be able to perform tasks that they would normally be unable to do without the use of such a system.

Over the last few decades, several projects have evolved which attempt to increase physician's effectiveness via the utilization of computers. There are far too many such projects to enumerate here. However, we will briefly consider a few to compare their approaches.

Expert systems such as MYCIN [11] have been developed to increase the efficiency of their users by allowing them to quickly consider the interrelationships between their available data. MYCIN is illustrative of a *rule-based expert system*, which allows the user to quickly run through a list of pre-defined rules which have been developed by experts in the field. These systems are of considerable value in well-defined, forward-chaining scenarios, where the particulars of the data are known (or suspected) to some degree and a conclusion, based upon known relationships, is desired.

In a 1979 study [17], a modified Turing test was employed to evaluate the quality of advice offered by the MYCIN expert system. A set of test cases were provided to expert clinicians and run through MYCIN; the resulting recommendations were judged by a panel of experts without knowledge of the source of the recommendation. The expert clinicians scored approval ratings as low as 42.5% while MYCIN scored a 65% approval rating. Furthermore, no human doctor was able to beat MYCIN. Schwartz et. al. [9] point out that, while "no other clinical expert system has been subjected to quite so rigorous a test, numerous studies, on a variety of different systems, have shown that expert systems are capable of exceeding [human] expert performance."

Knowledge-based computer systems have recently found their way into medical schools, as evidenced by the use INTERNIST-1 and ILIAD.

The INTERNIST-1/QMR [6] knowledge base has enjoyed successful integration into several medical school curriculums [5]. In its role as an aid the medical student, the system operates at three levels: as an electronic textbook; as an elementary spreadsheet for the exploration of diagnostic concepts, and as an expert consultant program capable of providing advice and direction to the user.

As an expert system, ILIAD [16] makes use of both Bayesian and Boolean *knowledge frames* in its representation of diseases encountered in internal medicine. Originally designed to aid medical students working in wards, the ILIAD system provides the user with differential diagnoses coupled with the ability to realistically simulate patient cases and accept textual data entry [4]. By virtue of its framing representation, the system is also capable of offering explanations to support its reasoning. In accordance with its design as an educational tool, ILIAD "allows free-text entry of observations made by the student during his/her workup of the patient, can provide consultation to the student at any stage of the process regarding differential diagnosis, and can advise the student on the most appropriate observation to make next [16]."

A different type of system is one which seeks to make available to the physician "everything" that is known (i.e., from journals, textbooks, reports, laboratory analysis, etc.) regarding a given problem at the moment that the information is needed. This concept of attempting to bring all the relevant knowledge on a given topic to bear on a patient's unique problem has gained more appeal with recent advances in telecommunications technology and has given rise to the concept of *problem-knowledge couplers* [14, 15]. Problem-knowledge couplers seek to:

> "...free the medical care provider of much of the impossible demand that he remember (or extract from the literature), judge the relevance of, and correctly apply the huge body of medical knowledge potentially available to him at the time of action [14]."

Beyond these "front end" advancements, which attempt to aid the physician at the point of diagnosis, there are also systems designed for "downstream" decision support. Such systems generally provide the user with two types of information: 1) data regarding the anticipated results and consequences of various treatment policies which enables the physician to prescribe the best possible treatment plan; and 2) a method by which to monitor a patient's treatment progress such that the dynamic, real-time adaptation of treatment is viable.

Many models already exist for statistical decision support in the form of formal hypothesis-testing algorithms and Bayesian updating procedures. Depending upon the scenario constructed, the physician is able to utilize standard decision analysis packages such as ARBORIST [12] and DAVID [10] or more sophisticated and dedicated diagnostic tools such as MEDAS [13], the latter being a Bayesian medical diagnostic system recognized as a leader among medical diagnostic programs. Unlike other medical expert systems, MEDAS combines expert system knowledge and relational database technology to: 1) permit the physician to maintain a current knowledge base without the need for specialized training; 2) exploit advances in pattern recognition technology during diagnostic inference and test selection; 3) utilize modern design methodology in the construction and accessing of its database; and 4), allow for the sharing of data between "independent" modules, thereby reducing the level of redundant data in an integrated system [13].

In addition, there is work presently being undertaken to not only increase the level of decision support available to a physician, but also to utilize the physicians' own comments, criticisms, and suggestions. One such system is DXplain [3], which was developed with the support and cooperation of the American Medical Association. DXplain can be thought of as a hybrid between an electronic medical text and an automated medical reference system which actively solicits the responses and criticisms of its users in order to accommodate their individual needs. It is able to provide thorough descriptions of over 2,000 different diseases, including symptoms, etiology, pathology, and prognosis. In addition, the system is able to provide up to 10 recent references deemed appropriate for each disease in its roster.

All these systems, though their methodologies and underlying features may differ, have one point in common: they attempt to bring as much knowledge and assistance as possible to the user at the precise time such knowledge and assistance is required, that is, "information on demand" at the point of use as compared to "current awareness" information as provided by journals, books, or lectures.

The Lure of Intelligent Systems

Despite the controversy surrounding their advent, medical intelligent systems are quite alluring and offer the contemporary medical professional several benefits never before available. The most commonly cited lures of intelligent systems for the medical community are:

Freedom from mundanity
Information integration and accessibility
Speed and accuracy
Heightened inferential power

For the moment, we will consider the first category, that of freedom. There is some debate as to the class of problems to which medical intelligent systems should be targeted.

There are some physicians who feel intelligent systems should focus on difficult cases where the doctor lacks training and/or experience. Others argue that computer systems should concern themselves with common/routine problems which the physician addresses more by policy than individual analysis, thereby freeing him to concentrate on more difficult cases. It is under this latter assumption that a newfound freedom from mundanity is perceived as a boon.

With developments in data structure, storage, and access, today's computers are well equipped to provide their users with the flexibility of data manipulation required for improved decision making and the capacity for scenario development and analysis.

Speed and accuracy are essential as the volume of patients and their flow rate through medical facilities increases. Computerization offers speedy data access/storage to address this problem and the concept of computerized medical records has gained a large following over recent years, both in terms of hospital operations and as an educational tool. By nature of its medium, electronic data frees the user of such tiresome tasks as deciphering handwriting and sifting through reams of papers for the required information. Not only is the speed of data retrieval increased, but also the data so retrieved can be of greater accuracy.

All of these concerns were addressed by one of the physicians interviewed when he described his "dream machine" to us:

> *"[What he] wanted was an Asimovian robot. He felt that his dream machine was a humanoid computer that would walk around with him, keep track of his ideas and appointments, issue reminders, be adept at document preparation, possess a built in telecommunication capability, and perform literature searches/analyses and database maintenance while he himself was asleep. The humanoid nature of the machine implies such features as ease of use, voice activation, and portability."*

Finally, we consider increased inferential power offered to the user. Once pertinent data is stored, it is not difficult to utilize software to display the data in a wide variety of forms (i.e., tables, correlated lists, graphs, etc.) and generate probabilistic scenarios. This ability to perform situation postulation and scenario analysis offers the user the ability to simultaneously consider several treatment options. In fact, one physician actually went so far as to describe his "dream machine" as

> *"... capable of predicting future events with probabilities (i.e., specifying scenario A will occur with probability p)."*

These contentions are supported by the physicians interviewed; our inventory of propositions is replete with statements such as:

- *Physicians have difficulty in obtaining medical records*
- *Intelligent systems should be able to access stored information*
- *More accessible information would be beneficial*
- *More accessible decision support would be beneficial*

Barriers to Acceptance

At this point one may ask: If intelligent systems are so wonderful, why aren't they used extensively by medical practitioners?

We have thus far identified three classes of barriers which severely limit intelligent system adoption and use by the medical community: techno-economic barriers, barriers to use, and cultural barriers.

Techno-economic barriers:

The concrete barriers to IS adoption and use are the most readily noticeable as they are directly observable and by no means subtle. The future is uncertain and many of the cost outlays over the next 5 to 10 years are being determined today.

> *"One physician was particularly frustrated over the planning of a new ambulatory care building to meet the needs for the year 2000 and beyond without possessing any knowledge of the state of health care even one year in advance. He was also very stressed over the need to plan staff additions without knowing the extent of health care reimbursements over the next few years.. He expressed his displeasure over a lack of up to date, on-line financial information."*

When considering an intelligent system, there is the initial question of compatibility to contend with -- will the proposed system be compatible with existing computer systems already in place? Will the proposed system require the purchase of additional technology? Such immediate concerns lead directly to the consideration of the proposed system's present and future hardware requirements, projected maintenance costs, and expandability. Many organizations severely undercount the necessary support features required for the effective running of a computer system.

While future intelligent system (IS) upgrades are expected, there is an underlying assumption that the actual hardware -- components, wiring, communication links, etc. -- are, with the possible exception of routine maintenance, constant over the life of the system. If one IS is purchased, it is assumed that another IS will not be required until the first has essentially "paid for itself." And yet, this assumption is not valid, especially when one considers, for example, the turmoil in the telecommunications industry over the advent of fiberoptic transmission capability.

In essence, the purchase of a good IS is presently perceived as a gamble The system obtained must be well suited to the practitioners' needs and budget over the foreseeable future. And yet, continuing advancements in computer technology have made new products obsolete almost as quickly as they are developed. Thus, the technical and economic barriers are quite formidable, especially to historically risk-averse institutions such as are found in the medical community.

Barriers to use:

The abstract barriers to IS adoption and use need not be as evident as the techno-economic barriers; nonetheless, they are just as formidable. Here the questions are not cost concerns, but rather functional considerations. Where do we get the data to use the system properly? How do we use the system properly? Do we need a devoted technician? Or two? Or three? Several propositions have rather pointedly noted these concerns:

- *Physicians don't know where to look or how to ask for decision support*
- *Physicians are concerned over the origins of the data used by decision support systems*
- *Physicians are dissatisfied by the necessity of middlemen* (i.e., technicians) *in their decision support systems*
- *The presence of middlemen* (i.e., technicians) *in the use of decision support systems blocks their use by physicians*
- *Some resistance to new medical technologies is due to the need for technicians to run the machines and the lack of communication and cooperation between the physicians and the technicians*

Simply obtaining the data to run can be a laborious task: two physicians can see the same patient and come up with totally different diagnoses and treatment plans. This leads to concern over the internal algorithms used by the IS to draw conclusions and just who is supplying the data used. This attitude was noted several times and formally expressed in an interview excerpt:

> *"...there are parts of the decision process we can't pin down; three different decision makers often arrive at three different conclusions with the same information."*

These considerations lead directly to the topic of system "transparency." There are two extremes: the glass box and the black box. Fully transparent ISs represent glass boxes. That is, the user can see exactly what the system is doing and how it is doing it. The other extreme is the black box, which magically spits out answers without apparent rhyme or reason. The primary argument against glass boxes is that the user simply doesn't need to know everything -- why bog the user down with complex algorithms or elaborate mathematical models? Too much transparency requires too much of the user's time and thereby reduces the user's efficiency. By the same token, getting an answer from a black box hardly invokes trust from the user. The IS must be designed, therefore, with enough transparency to allow the user to put his faith in the system without overburdening him with details.

The transparency issue leads into the subject of system technicians. In essence, present systems are often met with the complaint that they are so complex that hospitals require dedicated staff to take the physician's questions, translate them into a format that the system can understand, run the data, interpret the results, and give the physician his "answer." This cumbersome process quickly dissuades the physician from asking his question in all but the most difficult of cases. For instance, one physician interviewed reported that he was

> "... very frustrated with the poor accessibility of decision analysis (Bayesian) and support systems present in his work environment. The present system of decision analysis in his workplace requires the use of a middleman which is a block to its wide use; unfortunately, the middleman is required due to the complex nature of the software."

It is clear that one of the primary lures of ISs, that of speed, is lost. Further, the systems operating expenses are driven up by the salaries of the operators and the concrete barriers against new/improved ISs are given more weight as use of the present system declines. Finally, there is evidence to support a confusion in the minds of IS users between decision *making* and decision *support*. ISs offer their strengths in the area of decision support -- they exist, in other words, to *aid* the physician, not *substitute* form him. Too many users of ISs still feel that the system actually makes the decisions. This confusion is one of the major contributors to distrust in the so-called "black boxes."

Cultural Barriers:

The cultural barriers to IS adoption and use are arguably the most restricting and hardest to overcome. These problems are emotional and perceptual and hinge upon two issues: humanocentrism and discomfort.

Our interviews suggest that many physicians that use computers or ISs are quick to state that, no matter how good their system is, humans are better. They are convinced that no machine available today can perform any intellectual task better than a human being. To a large extent this is the result of professional training and self image, especially among physicians. It is arguable that a greater role of computers in medical school would dilute this aversion to ISs. Of those physicians interviewed, most had never touched a personal computer; some said that their training actually prohibited trust in computers. Consider the following propositions from our inventory:

- *Physicians are trained to think and operate as intelligent systems*
- *The vast majority of physicians do not plan on introducing a PC to their examining room*

- *Physicians believe that PCs shouldn't duplicate the work or thought processes of the practicing physician*
- *Physicians tend to avoid PCs because of little exposure in medical school and in their practice*
- *Physicians do not use PCs as assistants in their thought processes because they are trained to use their brains and distrust everyone else, let alone a machine.*

Issues such as uncertainty, distrust, and discomfort seem to be strong underlying factors in stalling the advent of ISs in the medical arena. There is a strong, almost debilitating, anxiety over failure in American professionals today; this anxiety is reflected more strongly in the medical community than in some other professions. Malpractice premiums are enormous and the number of lawsuits against medical practitioners is a strong deterrent currently, but may turn into a strong incentive for more reliance on technology of a "Best Available Technology" type. With so much riding on the correctness of his diagnosis and the results of his treatment plan, it is little wonder that many practicing physicians have not put their trust in untested intelligent machines, despite their increase in use of "hard technology" such as MRI, tomography, etc..

Compounding this apprehension over failure is the more subtle concern of obsolescence. Not for the obsolescence of the system per se (that is more a concrete barrier than a cultural issue) but for the physicians themselves. There is evidence to suggest that some physicians see themselves as the skilled laborer of yesteryear who wakes up one morning and finds out that he has been replaced by a machine. It may well be that this fear, under the guise of professionalism, is at the root of a great deal of resistance to the use of new medical intelligent systems.

Conclusion

There are an ever increasing number of reports that tout the superiority of certain systems over their human counterparts in diagnosing patients. Despite these apparent successes, physicians invariably stress that a *smart* machine that allows a practitioner to make better informed decisions is perceived as a great boon; however, there are far too many subtleties and nuances in medicine to allow for the development of a truly *intelligent* system. It is imperative that the human user be able to interject his own intelligence into the decision process if only to say "Okay, now I have enough information to make an intelligent decision."

At the same time, there is strong evidence of contention over just how helpful intelligent systems should be. It is questionable, at this time, just where skill *support* ends and skill *transfer* or *replacement* begins. Further, there is evidence to suggest that some physicians view the increased usage of early medical intelligent systems to be the result of a perceived decrease in the quality of new physicians. Consider the following propositions:

- *Medical intelligent systems will become more sophisticated and replace some physicians' thought processes because new physician's are of lower quality*
- *Medical intelligent systems will become more sophisticated and replace some physicians' thought processes because new physicians tend to transfer their professional burdens to machines*

The relationship hinted at here is that medical intelligent system usage may increase as physician competence decreases. The factuality of this allegation is less important than the existence of a perceived link between increased intelligent system usage and decreased user ability. Proper understanding of this link may prove invaluable to those interested in the proper product positioning of future medical intelligent systems and those planning medical education.

It is our belief that medical intelligent systems are best marketed in the near future as tools which increase a physician's efficiency and effectiveness, not machines which assume the physician's responsibilities or duties.

References:

[1] Bankowitz, RA., McNeil, MA., Challinor, SM., Parker, RC., Kapoor, WN., Miller, RA.: "A Computer-Assisted Medical Diagnostic Consultation Service." *Annals of Internal Medicine* (1989) 110, 824-32.
[2] Barnett, O.: "Computers in Medicine." *Journal of the American Medical Association* (1990) 263, 2631-33.
[3] Barnett, GO., Cimino, JJ., Hupp, JA, Hoffer, EP.: "DXplain: An Evolving Diagnostic Decision-Support System." *Journal of the American Medical Association* (1987) 258, 67-74.
[4] Bouhaddou, O., Warner, HR., Yu, H, Lincoln, MJ.: "The Knowledge Capabilities of the Vocabulary Component of a Medical Expert System." Presented at the Second Annual Chicago Medical Informatics Conference, Chicago, IL. (1990).
[5] Miller, RA., Masarie, FE.: "Use of the Quick Medical Reference (QMR) Program as a Tool for Medical Education." *Methods of Information in Medicine* (1989) 28, 340-45.
[6] Parker, RC., Miller, RA.: "Creation of Realistic Appearing Simulated Patient Cases Using the INTERNIST-1/QMR Knowledge Base and Interrelationship Properties of Manifestations." *Methods of Information in Medicine* (1989) 28, 346-51.
[7] Rubenstein, AH., Geisler, E.: "Users' Needs for Intelligent Support Systems (UNIS)

First Year Progress Report." CITT Document Number 91/02 (4/91) (1991).

[8] Rubenstein, AH., Geisler, E.: "Users' Needs for Intelligent Systems (UNIS): A Study of Potential Adoption by Professionals." Forthcoming in *Proceedings of the Workshop on Intelligent Workstations,* Siemens AG, Munich (1992).

[9] Schwartz, S., Griffin, T., Fox, J.: "Clinical Expert Systems Versus Linear Models: Do We Really Have To Choose?" *Behavioral Science* (1989) 34, 305-11.

[10] Shachter, R., Bertrand, L.: "DAVID: An Interactive Program for Processing Influence Diagrams, Version 1.1." Durham, NC: Duke University Center for Academic Computing (1988).

[11] Shortliffe, EH.: "Computer-Based Medical Consultation: MYCIN." New York: American Elsevier (1976).

[12] TEXAS INSTRUMENTS, INC.: "ARBORIST." Dallas, TX: Texas Instruments, Inc. (1986).

[13] Trace, M., Evens, M., Naeymi-Rad, F., Carmony, L.: "Medical Information Management: The MEDAS Approach." Presented at the Second Annual Chicago Medical Informatics Conference, Chicago, IL., (1990).

[14] Weed, CC.: "Problem-Knowledge Couplers." Concept paper written for PKC Corporation, South Burlington, VT, 05403, 1982. Presented at the Second Annual Chicago Medical Informatics Conference, Chicago, IL., (1990).

[15] Weed, LL.: "New Premises and New Tools for Medical Care and Medical Education." *Methods of Information in Medicine* (1989) 28, 207-14.

[16] Warner, HR., Haug, P., Bouhaddou, O., Lincoln, M., Warner, H., Sorenson, D., Williamson, JW., Fan, C.: "ILIAD as an Expert Consultant to Teach Differential Diagnosis." Presented at the Second Annual Chicago Medical Informatics Conference, Chicago, IL., (1990).

[17] Yu, VL., Buchanan, BG., Shortliffe, EH., Wraith, SM., Davis, R., Scott, AC., Gohen, SN.: "Evaluating the Performance of a Computer-Based Consultant." *Computer Programs in Biomedicine* (1979) 9, 95-102.

The Software Development Machine

Dean Koester

Siemens Gammasonics Inc.

Abstract

The software industry continues to suffer from cost overruns and delays due to poor understanding and management of the software life-cycle. This consumes a great deal of time and creates confusion for the engineer and others involved in development of software. The Software Development Machine (SDM) is an attempt to automate many of the areas of development that lead to these problems. The goal of its design is to provide life-cycle and configuration management, automated generation of requirements specifications, design documents, and code/applications throughout the development process, and automated software size estimation and schedule tracking. SDM would also allow for the capture and retrieval of knowledge and components for current or future use. Software quality is going to become a bigger issue in the industry as software is now involved in more and more critical areas of our life. Without an automated way of managing the many requirements and demands on the systems today, software will continue to take longer resulting in increased cost to companies and a decreased quality in software. This paper presents the objectives and components of the SDM in a potential working model which will enable management and engineering to produce higher quality products on time and within budget.

180

I. Introduction

After being involved in one of the largest Object-Oriented MacIntosh development efforts in history and interviewing engineers and managers, the dream machine for software development began to take shape. The Software Development Machine *(SDM)* may seem far fetched at first, but it can be obtained since many of the components already exist today. The Object-Oriented technology provides a great foundation on which to build the SDM. The real world is made of objects, and it is the mapping of these objects to the computer that make this technology so powerful. It is the natural language of the customer as well as the natural language of the software engineer.

" As computer professionals, we strive to build systems that are useful and that work; as software engineers, we are faced with the task of creating complex systems in the presence of scarce computing and human resources. Over the past several years, object-oriented technology has evolved in diverse segments of the computer sciences as a means of managing the complexity inherent in many different kinds of systems. The object model has proven to be a very powerful and unifying concept" [Booch 1991].

This paper attempts to identify the key components of an ideal machine which deals with the various complex tasks facing the software engineer/manager during the development of a large scale software system.

II. Problem

SDM attempts to solve several problems that exist in just about every software development effort today. The first is that there are major cost overruns and time delays due to poor software time estimation. Few companies today use any kind of metrics for estimating the size of a project. Because of this, projects are started with only a "finger in the wind" estimate and often are understaffed and ultimately over budget. The second problem is a lack of an automated way of developing an application that ensures meeting the customer's requirements. There is not an automated way of tracking these requirements all the way from the original software requirements specification into the code. The third problem is that there is not existing way of tapping into the knowledge of previously developed systems. In other words, many products are developed without tapping into how preceding and similar systems had been developed. As a result of this, many of the same errors are repeated. There is also a need to share knowledge and resources across divisions so the company as a whole can benefit. It has been my experience that no unifying tool exists to pull the whole software development life-cycle together. This absence creates frustration for the engineer and financial loss for the company.

III. Objective

The objective is to have a machine/process that aids the engineer in requirement analysis, estimation, design, documentation, verification/validation, and development. The machine must be modifiable to adapt to each unique development environment, so it will be used by all the engineers as well as other support personnel such as marketing and product reliability. The machine can produce various deliverables upon request such as a software requirements analysis, a software design specification, a detailed design specification, and estimates at any phase of the life-cycle with confidence level. The following are some of the major objectives of the SDM.

A. Automated estimation and schedule tracking at each phase of development.

It was our experience, after the initial plan was laid out and a crude estimate of the amount of work was made, the schedule and plans were never updated by the team. It would be a great benefit if the schedule could be continually updated automatically as the system obtains more data regarding the status and changes in the project plan.

B. Manages change and performs change impact analysis on schedule, resources, requirements, and existing code.

During the development of the project the team was constantly interrupted by marketing and other groups within the organization to make "small" changes to the requirements. Very often the engineer was asked how he thought this change might impact schedule without really having a good handle on how this change effected the entire design. We needed a way to trace this requirement change through the entire system to see exactly all the pieces it effected and how it effects the schedule with regards to both time and resources.

C. Configuration management for document and source control.

This may seem obvious, but no software development environment could function efficiently without a source and documentation control system. This could be enhanced by also keeping linkages between documents. For example: Requirements -> Object Design -> Design Review -> Code -> Code Review -> Test Plan -> Test Results. The configuration management system could also allow individuals to subscribe to any piece of information and to be notified when it has changed. It was our experience that often one document will reference another and is dependent on certain areas of that document to remain constant. If those areas change or are moved, the original document is no longer completely useful.

D. The ability to tap into already developed components as well as the ability to help identify potential reusable components.

One of most important attributes of the object-oriented technology is the potential for reuse within and outside the application. We definitely benefited greatly from reuse in our application, but we realize we could have done much better if we had recognized from the beginning other areas that had great potential for reuse. Other development teams at our company also use object-oriented technology with similar requirements with whom we could have shared a tremendous amount code, but we had no way of investigating what they were doing other then informal meetings. We needed a central depository of reusable components that was company wide.

E. Ability to support rapid prototyping and "what if" scenarios.

Often the customer does not know exactly what is wanted until a working application or prototype exists which he can critique. The sooner that all requirements are known the faster the engineers can analyze and design. "Remember that requirements analysis and design is like walking on water. It's easy when they are frozen."

F. Encourage a consistent methodology to be used in the analysis and design.

Another key issue that so often is ignored is that software development is much easier if everyone involved is communicating with the same language. In other words, it is very important for the development team to be using one methodology with which to communicate and validate their ideas/designs. In our project, we did not have any methodology chosen so each member had a different way of illustrating his ideas. This greatly slowed us down , since we all had to learn each others methodology (In most cases no methodology at all was used).

G. Accessibility by other departments such as marketing, product reliability and project management.

During the development of our system we were constantly asked what the status of the project was. If project management had access to a master schedule that was updated automatically as to the exact status of the project, they wouldn't have to depend on us to inform them (especially since we were never really sure beyond our particular area of interest anyway).

H.   Assists in analyzing and managing of requirements. (Aids decomposition and abstraction of requirements.)

There is a need for the automation of requirements analysis. Object-oriented technology is the mapping of real world objects into the computer. Currently, most engineers do this by hand. An engineer or a team of engineers attempt to assimilate all the requirements and abstract out common traits and behaviors of identified objects.

I.   Ability to be easily modified to meet new development requirements and methodologies (rules).

As new methodologies and technologies are developed, the engineer must have a tool that can evolve with these changes or it will be obsolete in a matter of months. This is especially true with regards to object-oriented technology, since it is still in its early stages of maturity.

J.   Reverse engineering for poorly documented code. (Generates design documents and requirements from already existing code.)

In order to understand and integrate code from other sources it may be necessary to reverse engineer. It may also be nice to reverse engineer systems from other platforms in order to then generate a new application for a new platform. For example take preexisting code for an IBM PC and reverse engineering it by creating the design and requirements specifications, now the engineer can generate the application for another platform or use pieces of it for another system.

K. Automated notification of changes and other types of communication

It was frequently our experience that a developer would change the behavior or interface of a particular object on which another developer was dependent. This caused either build or run time errors. It would be desirable to set up a publisher/subscriber mechanism of communication between developers who are dependent upon each others work. In other words, the developer who is dependent on the output of another would be notified and warned if any changes occur in the areas upon which he is dependent. Likewise the developer creating the change would be warned that the changes he is making is affecting others work.

L. Consistancy checking of requirements at every stage of the software process.

During our development it would have been very useful to periodically verify that each requirement was accounted for in the analysis, design, or coding phase of the project.

M. A feedback mechanism to drive continuing education.

The reality is that not all programmers are good object-oriented engineers. Currently there is no way to capture the decisions of the good object-oriented analyst/designer in order to teach other engineers.

This paper will address each of the above goals as reasonably obtainable goals. "Success in building software products means meeting customer needs, on a schedule responsive to the market, at the right cost, and with high quality" [Robertson, 1986] .

IV. The SDM Model

By providing five main services, the SDM answers each of the desired objectives. They are as follows:

A. Life-cycle and configuration management which ensures the linkages among specifications, requirements, code, test results, design documents, and any other needed information during the development cycle are maintained as well as allowing the flexibility to enter in and make changes at any point of the life-cycle.

With the onset of Object-Oriented technology, software engineers and managers have needed to make changes in the way they develop software. Previously, the "Waterfall" life-cycle was the dominant life-cycle approach to developing software.

" The waterfall life-cycle approach came into existence because of the grossly disorganized attempts to develop software that seemed all too common. You know the story: software is late, buggy, much too expensive, and ill-suited to the user's need. Programmers would "get the idea of what the client wanted," and then proceed to write a bunch of code. Once done coding, the paperwork (e.g. design) was attempted - but only until the next project started." [Berard 1991].

The waterfall life-cycle usually means that all the requirements analysis is complete before going into design, and all the design must be complete before coding. This life-cycle approach is no longer applicable for the object-oriented software development approach.

The SDM uses the "Recursive/Parallel" life-cycle approach to software development as a model of behavior. It basically allows the engineer more flexibility at all phases of development. The recursive/parallel approach simply allows the engineer to analyze an little, design a little, and code a little. In other words, this follows the natural way the engineer attacks large problems. It was our experience that not all requirements are at the same level of detail, and often some are not fully understood right away until some prototyping is done. Also, with an object-oriented approach to software development there is no obvious distinction between where requirements analysis ends and design starts. The main reason for this being that the job of the analyst is to map the real world requirements into the world of objects in the computer. While doing this mapping, the analyst starts making many design decisions regarding what objects are in the system and what behavior and attributes they have.

The following objectives are achieved by this service: C, G, K, L

B. Automated generation of requirements specifications, design documents, and code/applications from any point with in the life-cycle.

Many hours are spent by the engineer during the analysis and design phases of the project doing diagrams and changing these diagrams to reflect changes in requirements or design. One aspect that is often neglected by the engineer is to make sure that all documents are reflecting the current state of the project. For example, changes at the code level are almost never bubbled up to be reflected in the design documents if necessary. If the changes are guaranteed to be reflected at all levels of the life-cycle then the engineer can make changes at any point necessary.

The following objectives are achieved by this service: B, E, F, H

C. Automated schedule tracking with software size and quality metrics.

Engineers are typically very poor estimators, but companies are forced to count on these very inaccurate estimates for project budgeting and other such important decisions. If these estimates where done automatically based on a database of metrics that was constantly being updated as the system got actual results back, the company could get more reliable results as well as free the engineer up from this responsibility. There is also a great benefit to be gained if the project status was continually updated. This would allow management to monitor potential problem areas and make changes early to avoid

188

schedule slippages at the end of project. This is also beneficial in performing impact analysis if a partical requirement changes.

The following objectives are achieved by this service: A, B

D. Capture and retrieval of knowledge and components for current of later use.

Reuse can definitely be a great benefit for any project as well as across projects, but unless the tools are available with in the environment to query for potential reusable components as well as help the engineer construct these components then reuse probably will not happen. This service could even include the tracking of how and why certain decisions were made within the project life so others could learn from both good and bad decisions. Not all engineers make good object-oriented analysts, but by capturing why certain design decisions were made others in the company can benefit from the experts.

The following objectives are achieved by this service: D, J

E. Easily interchangable components to react to changes in methodology.

The SDM is also object-oriented in the sense that it is made up of components that can be customized to meet the current requirements of the project. For example, the code generating component could be replaced to produce PASCAL instead of C or generate the code for MS-DOS instead of UNIX.

The following objectives are achieved by this service: I, M

V. The SDM Model Level 1

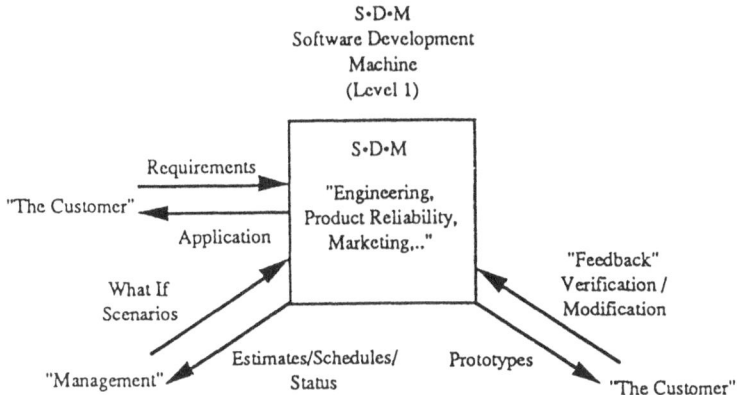

The software development machine (SDM) requires as input the customer requirements. From the requirements, the machine produces an application.

This is a high level model of the SDM (Level 1). The SDM takes in customer requirements in a specified format and from that generates an application. Many interactions take place along the way, but from the point of view of a customer (actual end user, marketing, product reliability, project management...) this is how the SDM would appear to them. One of the most important traits of the SDM is that it can be used by others besides engineering. Management can use it to obtain project status as well as obtain time and resource estimates for any set of requirements. Marketing can use it to assist customers in defining requirements. Prototypes can be used by the actual customer to give valuable feedback early in the project design phase. By allowing and encouraging other departments to be involved in the entire software development life-cycle, the others can be educated in some of the complex issues software engineers are faced with during the development. This tool may be just the vehicle to pull the entire development team together through out the life-cycle.

VI. The SDM Model Level 2

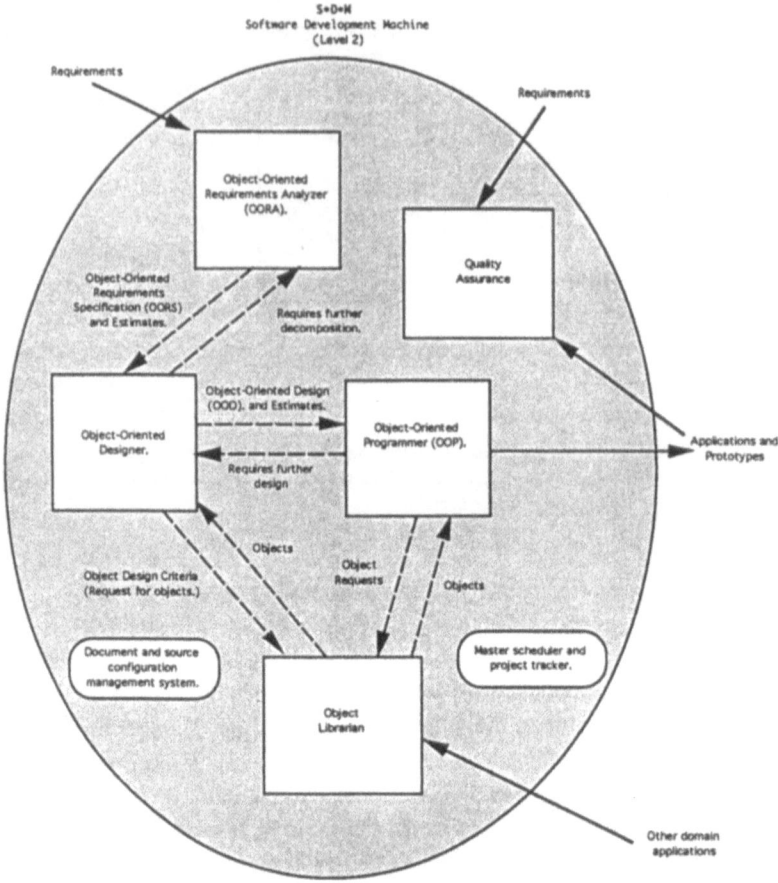

Before describing in more detail each component of the SDM, the following is a brief overview of the SDM at this level of decomposition. This model shows more detail of the proposed SDM. Each component of the diagram represents a stage in the breakdown of the requirements into an application. The SDM is an object-oriented abstration of the real world development process. Each component has well defined interfaces and responsibilities. This allows the replacement of the components without effecting the surrounding components. For example, if a different type of programming language or platform is required then the object-oriented programmer component is exchanged for the desired one. It maybe that initially many of this components require alot user interaction, but as the machine gains knowledge about the project and the environment many of these stages increase in the degree of automation.

The user of the machine can view the project at any perspective of the components. The machine manages the interconnections and linkages of the various outputs for the engineer. For example, the requirements for a particular object in the system can be traced back to the original requirements specification. This is especially useful when impact analysis needs to be performed on a system to determine the effect on the design/code by simply changing one requirement. This allows marketeers to play "what if" games with the requirements to determine the best combination of features to be added with minimal impact on time to market. Each component communicates to the master scheduler and project tracker as to current status so the master schedule can be updated. The software engineer is primarily interested in the object-oriented requirements analysis, the object-oriented designer, and the object-oriented programmer. These three work closely together to produce an application. At any time the engineer can choose to enter in at any component level. For example, after analyzing the requirements with the help of the analyzing component, he may chose to start designing a particular subsystem of the project. After designing this subsystem he then can prototype it with the use of the programming component that generates code and code templates from the design input. Each component of the SDM also interfaces to the document and source configuration management system to maintain linkages and version control.

A. The Object-Oriented Requirements Analyzer

This stage takes as input the set of customer requirements and abstracts objects out of them with its appropriate subset of requirements. The output of this stage is to produce an object-oriented requirements specification to be used by the object designer. This set of requirements is then tracked and verified continually by the SDM.

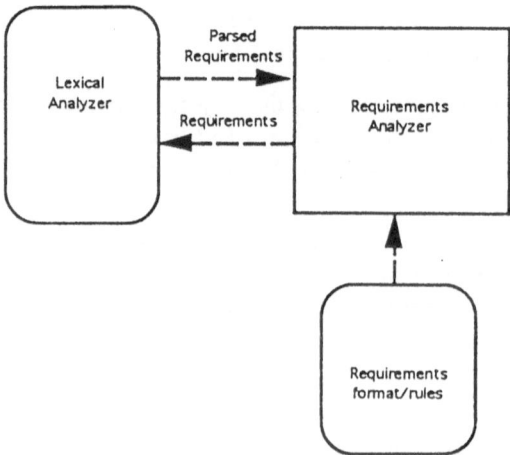

The requirements analyzer uses a lexical analyzer to parse the requirements into potential objects and their associated traits and behavior. This is then formated as to the required format for the object-oriented designer.

B. The Object-Oriented Designer

This stage understands object-oriented requirements as input. Given this input in combination with the object librarian's support, produces an object design for each object in the specification. The designer may need the requirements to further analyzed by the requirements analysts in order to continue with design. Since the designer can generate diagrams automatically it will be a great time saver for software designers and analyst.

C. The Object-Oriented Programmer

This stage understands the design format from the object-oriented designer. From the design document and the library of reusable software components it will produce an application. If the design is not detailed enough for the programmer it will request further design from the designer.

D. Quality Assurance

This box ensures that the resultant application meets the original set of requirements and that all requirements can be traced from the code back throughout the design to the requirements specification. It could also be used to ensure that all the objects in the system have been inspected and tested according to the designed unit test plan as well as according to the integration test plan.

D. Object Librarian

This manages the library of reusable software components. It is used to extract reusable components out of existing applications as well as aid the machine is design and production of applications by suggesting the use of already written software objects. Reusability metrics could be used to determine if an object should be place in the reusable object library. What is probably the most useful function the librarian can provide is the "object finder" service. This service searches for already exisiting components or components currently under construction that closely match the needs of the system. This search could go across departments, divisions, and even include other cooperating corporations. This may also serve as a class browser to allow the user to manually view the various classes available.

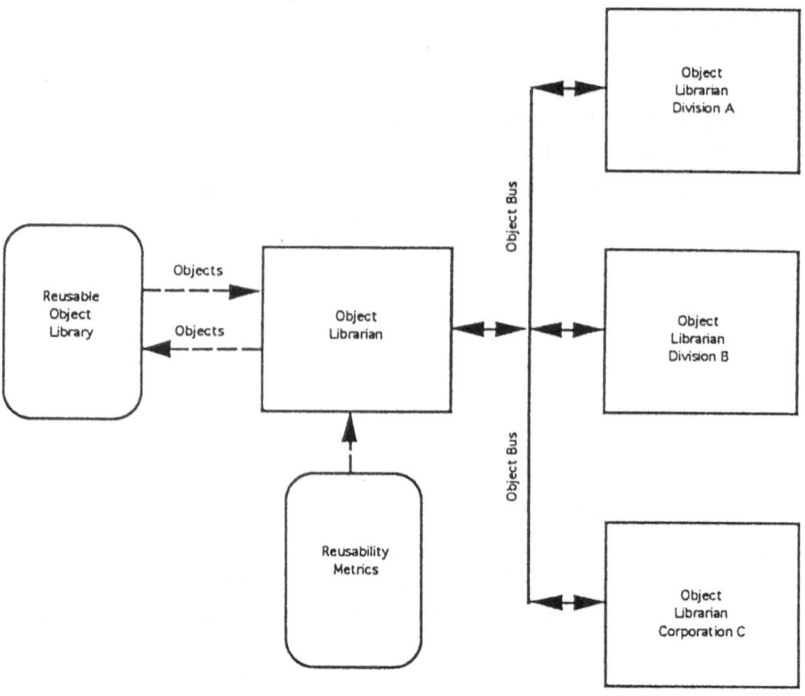

E. Document and source configuration management system

This maintains all the documents and code in the system as well as the linkages between them. This naturally includes version control and the ability to control access levels on files to allow only the desired people to have access to view or modify the document.

F. Master scheduler and project tracker

This component is constantly monitoring the progress of the output of each component and how it impacts the master schedule. This is used by others outside the software department such as project management to monitor the progress of the project. It may also be used for example, to determine the impact on schedule by changing requirements or design during the development of the system.

VII. Conclusion

The key objective of the SDM is to provide an integrated solution to managing the entire software development life-cycle from requirements acquisition and refinement to maintenance. There are many possible building blocks for which to create the SDM. There are many products available to address the individual components of the SDM, but there is not one unifying solution that encompasses all aspects of the software engineer's job. The SDM stresses that software is no longer localized to a few individuals but now involves many departments within the corporation and has had major impact on many company's bottom line. Software development can no longer be treated with disrespect. The SDM should be viewed not just as an idea, but as a necessity if the software industry is to keep up with other technologies and the increasing demands customers. The software industry is at a very critical turning point since we have not evolved fast enough to keep up with the many demands of the market place, but the company who can master the software life-cycle will be successful. The crisis of managing complexity is not localized to software development, but is interwoven in just about every department. The SDM could even serve as model/abstraction for others to utilitize.

VII. References

Abdel-Hamidm Tarek K., and Madnick, Stuart E.

Impact of Schedule Estimation on Software Project Behavior, IEEE Software, July 1986, pp. 70-75.

Arango, Guillermo

A Tool Shell for Tracking Design Decisions, IEEE Software, March 1991, pp. 75-83.

196

Barnes, Bruce H. and
Bollinger, Terry B.

Making Reuse Cost-Effective, IEEE Software,
January 1991, pp. 13-24.

Berard, Edward V.

Life-Cycle Approaches, Berard Software
Engineering, Inc., January 1991.

Boehm, Barry W.

*Software Risk Management: Principles and
Practices,* IEEE Software, January 1991, pp.
32-41.

Booch, Grady

"Object-Oriented Design with Applications",
Benjamin/Cummings Publishing Company,
Redwood City, CA., 1991.

Coad, Peter J.

Why use object-oriented development?,
Journal of Object-Oriented Programming,
October 1991, pp. 60-61.

Coad, Peter J. and
Yourdan, Edward

"Object-Oriented Analysis", Yourdon Press,
Englewood Cliffs, New Jersey, 1990.

Genuchten, Michiel van

*Why is Software Late? An Empirical Study
of Reasons for Delay in Software
Development,* IEEE Transactions On
Software Engineering, Vol. 17, NO. 6,
June 1991, pp. 582-590.

Humphrey, Watts S.,
Snyder Terry R. and
Willis Ronald R.

*Software Process Improvement at Hughes
Aircraft,* IEEE Software, July 1991,
pp. 11-23.

Kraut, Robert E. and
Streeter, Lynn A.

*Coordination in Large Scale Software
Development,* Bellcore, Morristown, New
Jersey, 1990.

Laranjeira, Luiz A.

Software Size Estimation of Object-Oriented Systems, IEEE Transactions On Software Engineering, Vol. 16, NO. 5, May 1990, pp. 510-521.

Lee, Sangbum

Object-Oriented analysis and specification:, Journal of Object-Oriented Programming, January 1991, pp. 35-42.

Levendel, Ytzhak

Improving Quality with a Manufacturing Process, IEEE Software, March 1991, pp. 13-25.

Miriyala, Kanth and Harandi, Mehdi T.

Automatic Derivation of Formal Software Specifications from Informal Descriptions, IEEE Transactions On Software Engineering, Vol. 17, NO. 10, October 1991, pp. 1126-1142.

Moriconi, Mark and Winkler, Timothy C.

Approximate Reasoning About the Semantic Effects of Program Changes, IEEE Transactions On Software Engineering, Vol. 16, NO. 9, January 1990, pp. 980-992.

Pressman, Rodger S.

"Software Engineering A Practitioner's Approach", McGraw-Hill, New York, NY., 1987.

Protsko, Beth L., Sorenson, Paul G., Tremplay, Paul J., and Schaefer, Douglas A.

Automatic Derivation of Formal Software Specifications from Informal Descriptions, IEEE Transactions On Software Engineering, Vol. 17, NO. 1, January 1991, pp. 10-21.

Rammamoorthy, C.V., Usuda, Yutaka, Prakash, Atul, and Tsai, W.T.

The Evolution Support Environment System, IEEE Transactions On Software Engineering, Vol. 16, NO. 11, November 1990, pp. 1225-1234.

Roberston, Lenard B.

Effective Management of Software Development, AT&T Technical Journal, Vol. 65, No. 2, March/April 1986, pp. 94-101.

Soni, Dilip

Intelligent Support for Software Maintenance, Siemens Review, Spring 1991, pp. 14-18.

Programming by Demonstration:
A Basis for
Auto-Customizable Workstation Software

Michael Sassin, Siegfried Bocionek

Siemens AG

Abstract:

Professional software developers' and users' requirements are difficult to satisfy by conventional tools, due to the individual's idiosyncrasies. One solution is to provide basic tools which can be adapted to specific wishes. Programming by Demonstration (PbD) is a powerful paradigm for fast software development that is able to support the development of auto-customizable software. The auto-customizing mechanisms are used to define new functions according to the user's preferences. New functions are created by demonstrating the desired functionality instead of explicitly encoding them.

1. Introduction

Meeting programming standards is one of the important requirements that software tool developers have to fulfil. This, however, leads to problems with the utilization of the product, since the users' requirements cannot be completely satisfied by a standard tool or environment due to his idiosyncrasies.

In practice, customizing is used to reach this aim. But the two currently used approaches now have problems that limitate the application in most cases:
1. One form of customization allows users to code new functions by means of a programming language. For example, every LISP environment can be extended by writing additional LISP functions. But software designers cannot assume that each user is able to program new features with a special programming language without any knowledge about the environment.
2. Customization using configuration tables is an approach that is more practical for users because they only have to select the options they want from a finite set of possibile alternatives. Nevertheless, this method is only sufficient for limited

customization and not for supporting particular users' wishes that had not been taken into account by software developers from the beginning.

One practical solution, in our opinion, is to offer a general mechanism that enables software users to create supplementary functions while using a tool. Such an *auto-customizing mechanism* has advantages for software developers and tool users as well:

1. Software users are able to adapt software to their individual wishes and preferences without needing knowledge about the environment used or a specific programming language. They can tailor the basic functions of the software according to their idiosyncrasies or create new functions if necessary.

2. Software developers can reduce development time because a smaller set of basic functions has to be programmed. Additionally, software developers themselves can use the auto-customizing mechanism to create further functionality without explicit programming.

Programming by Demonstration (PbD) is an approach for fast software development that supports the solution described above. Instead of explicitly encoding new functions, they are created by demonstrating the desired functionality. Therefore, even computer novices are able to create new functions by demonstrating the new application as a sequence of basic functions.

The PbD system traces actions during the manipulation of objects. These traces are analyzed and semantics are extracted. The result is represented as a function schema that describes the manipulation of objects. During the traces' analysis - in one step - the types of objects are generalized and the newly created functions are applied to them. Substituting constants by variables is another generalization example using a similar mechanism.

The new function is created "as general as possible". For example, in a graphics editor, the user can demonstrate how to extend a specific rectangle to a 3D box. The result is a function that is able to extend all closed convex polygons to 3D (see section 4).

The remainder of the paper is organized as follows: Section 2 gives an overview of related work. In section 3 we describe the requirements, concept and formalism of PbD and an architecture for auto-customizable software. As an example, in section 4, we show how a "2D graphics editor" can be extended in order to provide some 3D functionalities by adding new functions using the auto-customizing technique. In section 5 we discuss general problems with our PbD approach as well as particular difficulties of the graphics editor. Section 6 concludes the paper and outlines future research.

2. Related Work

PbD in Robotics

The APO project (Assembly Plan from Observation) [Ike91] is one of the first applications of PbD in the field of robotics. A human, solving an assembly task in a block-world, is observed by a video camera. From the observations, code is generated which enables a manipulator to repeat the task. However, the human interaction, esp. the hand movements, is not taken into account (except for start-/stop signalling). Furthermore, no other sensors are used, esp. no forces or torques applied to the objects are measured. In our opinion, the observation of those data is necessary for complex assembly tasks.

The project "Teaching by Showing" [Kun90] shows a similar application of PbD. It is also used for direct repetition in the field of robotics. A human solving a real world problem is observed by a camera. The observed event is automatically partitioned into basic actions that can be repeated by a manipulator. On top of these basic actions program code for a manipulator can be generated that repeats the demonstrated task. Not all problems, such as the recognition of basic actions or the uncertainty of the visual system, are solved. Nevertheless, the project "Teaching by Showing" is a promising example of PbD. It shows the power of this method and the problems which are to be expected.

PbD in User Interface Programming

The dissertation of B. Myers [Mye87] describes a programming system PERIDOT that allows the development of Macintosh-like user interfaces by demonstration. Scroll bars, menus, window layouts, and mouse interactions are "programmed" only by arranging graphical symbols and answering questions of the system. Recent work of B. Myers in the GARNET project [Mye90a, Mye90b] include new ideas and cover additional application areas such as text formatting, scientific visualization, business graphics, and computer games. This work seems the closest to our objective in the PbD exploration field.

Another good example of PbD in the graphics editor domain is the METAMOUSE [Mau89]. The Metamouse system traces the user's interactions, predicts what he will do next, and then automatically performs this action. If the user does not like it, he can cancel it by one mouse click.

An approach similar to Metamouse is the EAGER system [Cyp91], which is used in a Macintosh HyperCard environment. Like Metamouse, Eager monitors the actions of the user and anticipates from some observed regularities which action will come next. Then, Eager highlights all buttons and card fields that are related to the anticipated action. If the user wants Eager to perform the task automatically, he just clicks the Eager "cat head". If he clicks any other field, the Eager's guess is implicitly rejected. Eager performs well on infering loops from examples, yet nested loops and conditionals are still missing [Cyp91].

In constrast to Metamouse and Eager, we do not want to anticipate actions during normal operation of a tool. Instead, we will provide two modes, normal and demonstration (see section 3.3).

Learning from Examples

A topic related to PbD is *"Learning from Examples"* [Die86]. It is an approach for predicting the next event in a sequence. As an application of Learning from Examples, the card game Eleusis is presented. It tries to automatically detect the hidden rules of the game in order to predict the next card that will be played. The system contains different kinds of generalization functions to interpret a sequence of events.

Another important method for extracting the semantics of an event is *"Explanation-Based Learning"* (EBL) [DeJ86a, DeJ86b] Every new fact added to a knowledge base is explained based on a predefined domain. This explanation provides an effective mechanism for deciding whether a fact is an instance of a chunk or schema of the knowledge base. This approach seems very promising for the analysis of observed system behaviours. With the application of EBL techniques the interpreted events demonstrated by PbD can be analyzed and transformed into a program schema. Additionally, this technique is useful to extract recursive or iterative structures of the interpreted events as shown in [Sha89].

The Learning by Watching (LEW) project [Con90] is aimed at supporting data acquisition in a data base. By generalizing the differences between question/answer pairs, the LEW system learns to answer an analogously structured query. LEW's method of integrating negative examples is of special interest to our PbD work.

Learning by Analogy, Cased-Based Learning

Based on earlier work on learning by analogy [Car82] the PRODIGY project develops a planner with an integrated case-based learning component [Car89] and EBL mechanisms [Min89]. PRODIGY administrates a large case database of successful goal/plan pairs and, for every new goal, tries to find the most similar one in the database (case retrieval). Case replay then reduces to the modification of the existing goal/plan pair in order to meet the new goal [Vel92].

This approach may be helpful for our PbD approach because it allows us to make use of "experience", instead of learning new functionality only from the new examples presented.

In order to efficiently use the analogy and generalization techniques, systems have to provide mechanisms that reduce the number of facts or schemata containing the knowledge. The aim is to find a normal representation in order to delete all redundancies. The semantics of the representation have to be unchanged whereas the syntax can be changed. Such a structural generalization is shown in [Kod86] and mentioned in [DeJ86a].

Rewriting rules perform the transformations on the knowledge representation. This work seems to be of importance to our PbD research, especially as a way of simplifying the representation of the recorded events (see sections 3 and 4).

Program Schemata as a Representation Mechanism
Learning new functions by generalization always means finding some kind of new program schema. The cliché concept of the programmer's apprentice project [Ric88] is such an approach. It aims at enhancing the programmer's productivity by automatically providing preprogrammed code for standard tasks. [Nis91] also chooses this method. (We, in contrast, want to *generate* this code. However, the cliché structure seems to be a possible representation mechanism for us.)

3. Concepts

3.1 Concepts of PbD
Fig. 1 shows the principal structure of PbD applications. The demonstrated event has to be transformed into a fixed syntactic form, which is called *event description*. For every application domain a syntax of events has to be provided. It can be a trace of keyboard or mouse strokes in the case of a graphics editor, or a complicated "movement and grasping language" for robotics problems (for a first approach see [Kan91]). A "sentence" in such a language is derived from a vision system or other sensors such as the data glove and data suit.

A set of expert-supplied *property functions* is used to find the "meaning" of an event, called the *event interpretation*. Finding the "meaning" of an event is equivalent to the extraction of its semantics. In the case of a graphics editor the property functions extract geometric and spatial information such as angles, lengths, orientations, crossings or the relationship of all objects in the working window. In the case of a text editor, property functions can be defined e.g. as the length of a string, the value of the character under the cursor or a specific cursor position. The result of the interpretation is called the *event representation*. It is composed of the conjunct of all valid, instantiated property functions applied to the event description (which can be represented as a feature vector) and the event description itself. Consequently, the event representation contains the complete, in respect to the set of property functions, semantic and syntactic information of an event.

The generalization function is used to extract characteristic features from a collection of event representations. This results in a *function schema* that represents the control and "action" flow of the demonstrated events. The control part of the schema is derived from the properties of the event, whereas the "action" part is extracted from the action trace. As

204

a specification formalism for the schemata we will use an existing high-level language (such as CIP-L, [Bau85]).

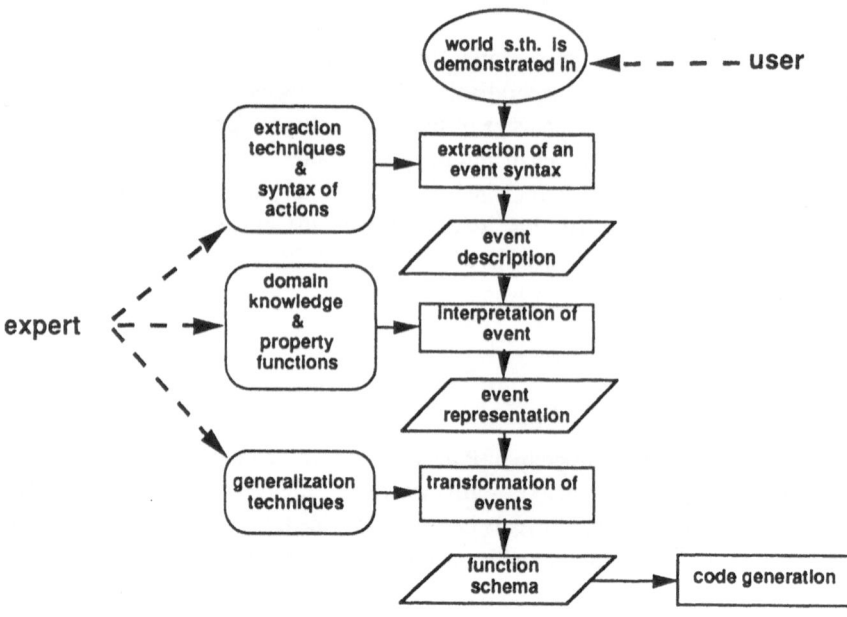

Fig. 1: Concept of PbD

We discuss two methods for finding the function schema. The first one is to partition the event representation such that loops are detected. For example, if the properties of the world states at the beginning of a successive set of partition hold the same constraints, a loop is detected. In general, these partitions can be used to derive schemata encapsulating basic actions e.g. "if-then-else" or "while" schemata. An instance of such a method is induction [Sha81, Mic83].

The second method entails to applying "Explanation-Based Learning" methods [DeJ86a, Sha89] to a set of event representations. It is especially important when deriving iterative and/or recursive structures and explanations.

In the last step the program code is generated that is suitable for all instances of the function schema. This should be done by the automatic derivation mechanism of the high-level specification language.

3.2 Auto-Customizable SW Tools Requirements

Providing auto-customizable software that enables optimal support for all software users, including computer novices, is an important precondition for intelligent workstations. In order to satisfy users and developers of auto-customizable software some requirements have to be fulfilled.

1. *Menu support*: The tool user should be provided with a menu-driven demonstration system because it is an environment that requires no deeper understanding of computers and is easy to use.

2. *Interactive guidance of derivation:* The interpretation of events and the derivation of generalized function schemata should be supported by question/answer dialogues between the PbD system and the PbD user. This helps to reduce search spaces, to select between more than one possible derivation and to request missing parameters.

3. *Explanation facility, UNDO*: The explanation facility should support the user on different levels such as testwise application of new functions, presentation of the formal derivation etc. An UNDO mechanism should allow him to reset immediately when unwanted results occur.

4. *Development support*: Efficient support for the programming phase is a significant facility for software developers. These mechanisms should be offered by means of suitable acquisition facilities for domain-dependent information, such as a specific world description, extraction techniques, property functions and generalization strategies.

5. *PbD interpretation mechanism*: An interpretation mechanism for the domain-independent PbD basics should be provided. Such an interpreter can be used by software developers to extend their basic software product by means of auto-customizing. It has to follow the top-level loop: record example - interpret example - generalize over a collection of interpreted examples - select and instantiate a function schema.

6. *"Strong" set of property functions*: During the development of an auto-customizable software product, programmers have to ensure that the set of property functions is "strong". This means, that the set of property function is "general enough" to extract all necessary information from the demonstrated example. This will enable the user to create functions that had not been taken into account by software developers.

206

3.3 Architecture of an Auto-Customizable SW Tool

Fig. 2 shows the architecture of an auto-customizable software tool. On top of a basic software tool (e.g. a DRAW program), an auto-customizable tool is constructed. It is composed of the slightly modified basic tool itself, a programming system that provides a menu environment, an extraction module, and a domain-independent interpreter that applies the domain-dependent information to the examples. In addition, the PbD shell provides software developers with the means for acquiring the domain-dependent information.

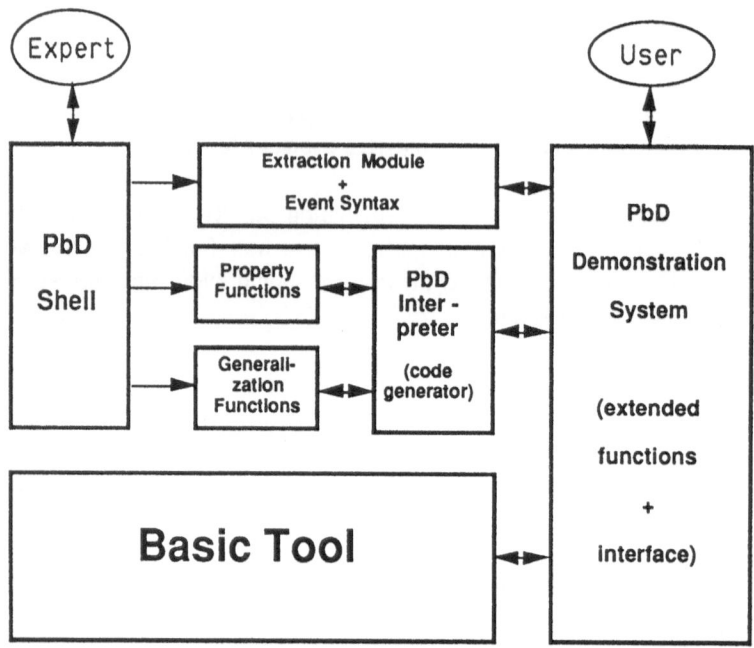

Fig. 2: Architecture of an Auto-Customizable Tool

The PbD shell is not usually given to tool users. However, to extend domain knowledge, experienced users should have the opportunity to assume the "expert's" role when neccessary.

The PbD demonstration system consists of a menu-oriented desk top that supports the demonstration and management of examples, the application of created functions and the management of messages and information such as error messages or information about generalization (see fig. 3). This also includes the guidance of questions and answers and all other explanation activities. Furthermore, it coordinates the communication between the modules.

The use of and the data flow in the auto-customizable tool can be described as follows. If the tool is in the normal mode of operation, the user can directly start the basic tool's functions. If the tool is in the demonstration mode, the event and the world sequences recorded during the demonstration are transfered from the basic tool to the extraction module. There, the event description is created and stored.

The process of interpretation, generalization and code generation is started by the user after a sufficient set of examples was demonstrated. The event descriptions of these examples are transfered to the PbD interpreter where new function code is generated and transfered back to the PbD demonstration system. There, the user can integrate the new function into the menu environment.

The extraction module and the PbD interpreter work as described in section 3.1.

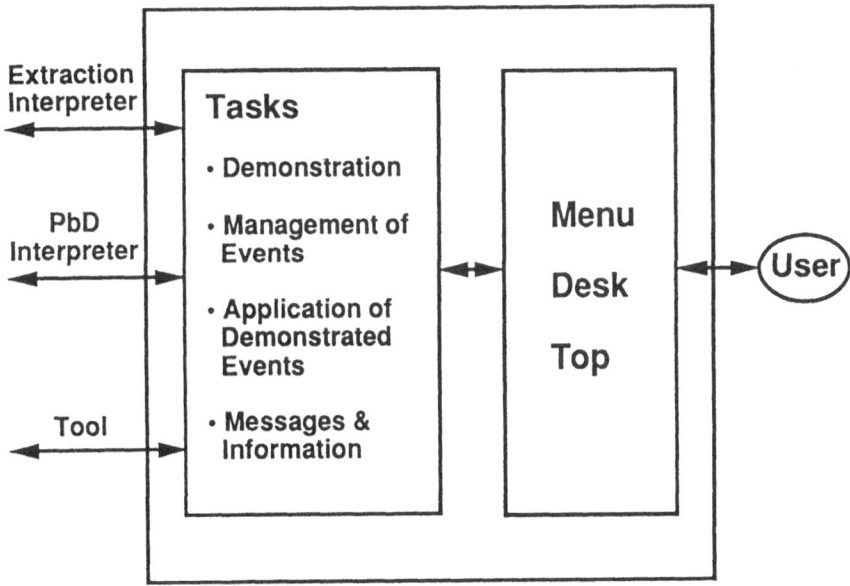

Fig. 3: PbD Demonstration System

The PbD shell is an efficient facility for the acquisition of domain-dependent information, such as the event syntax, property functions and generalization functions. This knowledge is formulated in a suitable modelling language. The PbD shell can be a simple text editor, but also a complex compiler.

4. Example: PbD as Part of a Graphics Editor

In this section we want to outline a scenario every user of DRAW-like graphics editors is familiar with. The problem sketched here is the lack of certain functions in such editors which, nevertheless, are frequently needed for the preparation of scientific papers and transparencies. One of those functions is the capability of extending some 2D objects to 3D objects in order to provide more "plasticity" in the drawings. Such tasks can, of course, be done by means of the editor's basic function set. But it is a boring and time-consuming job to construct 3D objects only having *drawline* and *copy* commands at hand. Buying a CAD package for comfortable 3D drawings, however, can be very expensive (for both, SW and HW), can take much time to master the tool and can cause compatibility problems when one wants to integrate the pictures in documents. Therefore, a DRAW editor with PbD capabilities, i.e. macro mechanisms plus generalization, may be the right choice to support comfortable document preparation by means of a self-defined additional functionality. The following sections shall illustrate how such a "Auto-Custom-Draw" package could work.

4.1 Providing the Domain Theory

As stated above, our PbD approach is a combination of domain theory and induction. The domain theory consists of the property functions together with a set of derivation rules to determine all (necessary) features of an example observed. These functions and rules belong to the domain-dependent part of the PbD enhancement and have to be defined by the "expert" in advance (see fig. 1).
As one possibility, all objects of the domain "Graphics Editor" are classified along a type hierarchy. Fig. 4, for example, shows a tree-like classification that divides all objects in closed and open ones, the former in closed rounded objects and closed polygons, the closed polygons in concave and convex ones, and so forth.

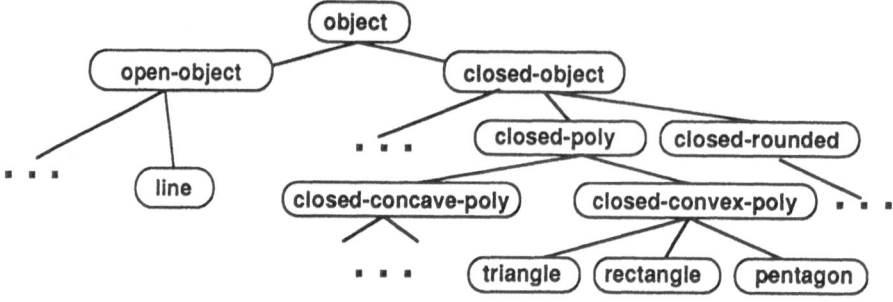

Fig. 4: Example of a Type Hierarchy

In a PROLOG-like notion we can describe the contents of the graphics screen through facts like

```
in-scene (rectangle, a).
```

expressing that a rectangle with name *a* is part of the current scene. By rules like

```
in-scene (A,B) :- is-a (C,A), in-scene (C,B).
```

and facts like

```
is-a (rectangle, closed-convex-poly).
```

we can model the type hierarchy for our domain and derive facts like

```
in-scene (closed-convex-poly, a).
```

saying that our rectangle *a* in the scene is also of type closed-convex-poly. Later on we will need this mechanism of derived types for the generalization of objects with different types.

Object Properties

The property functions for graphical objects, which describe their basic geometric attributes, are also part of the domain theory. In case of a line *l*, for example, one would define

```
length (l, 125).
gradient (l, 320).
startpoint (l, p1).
endpoint (l, p2).
```

and the properties for points in the workspace as

```
coordinates (p1, 27.5, -22.0).
```

Object Relations

Object relations are property functions that (among other possibilities) describe the spatial arrangement of the objects on the screen. They can be organized in form of decision trees such as that in fig. 5.

Fig. 5 states that, for example, a line *l* can be related to a rectangle *r*, that this relation can be *l crosses r*, *l lies inside r* or *l lies outside r*. Lying outside may have two meanings: *l avoids r* (*parallel* to one of *r*'s sides or *inclined*), or *l meets r* at exactly one point (*at a corner* or *at any point* of one side). The situation in fig. 6 can, in our domain theory, be described as a so-called *feature vector*:

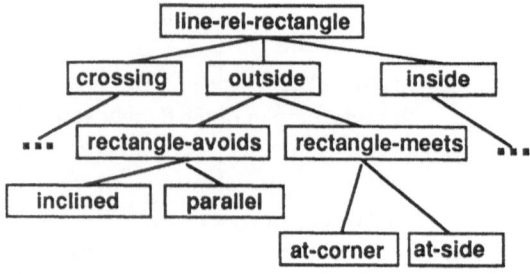

Fig. 5: Object Relations between Lines and Rectangles

```
in-scene (line, l).
in-scene (rectangle, r).
line-rel-rectangle (l, r, outside).
line-rect-outside (l, r, meets).
line-rect-outside-meets (l, r, at-corner).
line-rect-outside-meets-at-corner (l, r, p1).
```

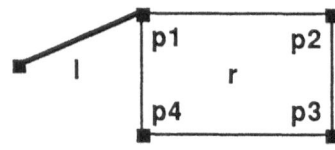

Fig. 6: Line l meets Rectangle r outside at the Corner pl

All facts describing object relations are derived by using rules, the object properties and other facts that were derived at any previous step.

Object Relation Properties
Object relation properties are property functions that add specific attributes to object relations. For example

```
distance-of-parallels (l1, l2, 200).
```
states that two parallel lines *l1* and *l2* are *200* units distant from each other. This result can be derived by applying the rules

```
parallel-lines (X, Y) :-
   is-a (X, line), is-a (Y, line), X neq Y,
            gradient (X, G), gradient (Y, G).
distance-of parallels (X, Y, D) :-
   parallel-lines (X, Y), calculate-line-distance (X, Y, D)
```

and providing another rule or function *calculate-line-distance* to compute the distance between two lines from the coordinates of their start end endpoints.

4.2 Interpretation and Generalization

Interpretation means that the property functions, as defined in the domain theory, are applied to the objects on the screen and to the actions performed by the user. Fig. 7 shows a sequence of four world states (= subsequent screens) w_0 to w_3. The sequence is intended to demonstrate to the system how to draw the supporting lines when extending a rectangle to a 3D box.

Fig. 7: Demonstration of Supporting Lines for a 3D Rectangle

During the *generalization* step the PbD interpreter compares the feature vectors of all world states (= examples) and finds out the common properties according to *domain-dependent strategies*. Those strategies heavily depend on the function schema that shall finally be synthesized from the generalized examples. Here, the following strategy is chosen: The feature vector of w0 becomes the precondition Pre of a schema

```
IF   Pre   THEN   REPEAT   drawline (X)   WITH   Post (X);
```

Post (X) is the postcondition that every line X must fulfill. It is yielded by the generalization of the feature vectors of the world state w_0 to w_3 Then, the synthesized instance of that function schema looks like

212

$$\text{IF} \begin{bmatrix} \texttt{in-scene (rectangle, r)} \\ \texttt{length (r, la)} \\ \texttt{width (r, ba)} \end{bmatrix} \textbf{THEN} \texttt{ drawline (X) } \textbf{WITH}$$

$$\begin{bmatrix} \texttt{length (X, l)} \\ \texttt{gradient (X, g)} \\ \texttt{line-rect-outside-meets(X, r, at-corner)} \end{bmatrix}$$

Here, the PbD system still "believes" that the rectangle r, with a specific length *la* and width *ba*, is important for the precondition. This can be generalized by demonstrating examples of rectangles (already extended by supporting lines) with different size, position and orientation as in fig. 8.

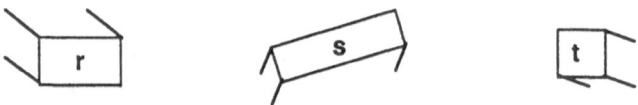

Fig. 8: Generalization over Different Rectangles

The result of the generalization is that the fixed name of rectangle *r* in the function schema's precondition is replaced by a variable *Y*.

$$\textbf{IF} \texttt{ [in-scene (rectangle, Y)] } \textbf{THEN} \texttt{ drawline (X) } \textbf{WITH}$$

$$\begin{bmatrix} \texttt{length (X, l)} \\ \texttt{gradient (X, g)} \\ \texttt{line-rect-outside-meets(X, r, at-corner)} \end{bmatrix}$$

When, as in fig. 9, triangles, pentagons and other closed polygons are demonstrated the generalization changes all occurencies of types in the synthesized schema to *closed-poly*. This is achieved through the type hierarchy because the generalization algorithm always searches for the "lowest common type" in the type tree if objects with common properties differ only in their type.

Fig. 9: Generalization of Types

The function schema belonging to fig. 9 is

IF[in-scene (closed-convex-poly, Y)] **THEN** drawline(X) **WITH**

$$\begin{bmatrix} \text{length (X, 1)} \\ \text{gradient (X, g)} \\ \text{line-outside-poly-meets-at-corner(X, Y)} \end{bmatrix}$$

After having demonstrated the necessary examples to draw the supporting lines of all closed polygons in order to extend them to 3D we have, now, to demonstrate the construction of the closing lines. Fig. 10 shows the sequence to learn that task for one specific rectangle *r*.

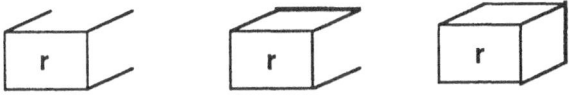

Fig. 10: Demonstration of the Closing Lines

The instantiated function schema to draw all closing lines is then

IF
$$\begin{bmatrix} \text{in-scene (rectangle, r)} \\ \text{in-scene (line, P1)} \\ \text{in-scene (line, P2)} \\ \text{co-connected (P1, r)} \\ \text{co-connected (P2, r)} \\ \text{par-neighbor (P1, P2)} \\ \text{length (r, 12)} \quad \cdots \end{bmatrix}$$
THEN drawline (X) **WITH**
$$\begin{bmatrix} \text{c-connected (X, P1)} \\ \text{c-connected (X, P2)} \\ \text{outside-parallel (X, r)} \end{bmatrix}$$

"co-connected" is a derived property expression that a line is connected to a polygon's corner. "par-neighbor" is a derived property. It expresses that that line P1 and P2 are paralled and that no parallel line lies inbetween. "c-connected" describes that a line X is connected to one endpoint of a line P1 (and P2 resp., in the example).

Again, by demonstrating the same task with different rectangles we can introduce the variable *Y* to substitute the fixed rectangle *r* in the precondition. However, when repeating the same step with other closed polygons we will fail when drawing all closing lines of a concave polygon like that in fig. 11.

214

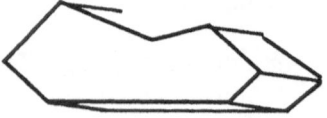

Fig. 11: Constructing Closing lines for Concave Polygon Fails

One of the closing lines needed cannot be drawn because the endpoint would lie inside the polygon and, hence, drawing the line would violate the postcondition that all closing lines have to lie outside the polygon. The property functions - as described so far - do not allow the PbD interpreter to find a concept for a hidden line algorithm in general (see the discussion below).

As the result, the combination of the complete drawing series, we have still to demonstrate the combination of *draw-supporting-lines* and *draw-closing-lines* to complete the originally required function *extend-object-to-3D* (see fig. 12). This is possible in a simple macro-defining style. Note, that the precondition in *draw-supporting-lines* has to be reduced to *closed-convex-poly* instead of all *closed-polys*.

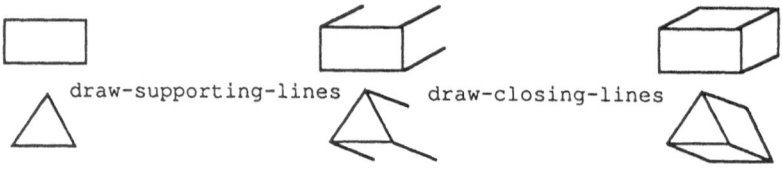

Fig. 12: Definition of Macro "extend-object-to 3D"

5. Discussion

There are many different problems to cope with when trying to make the PbD approach really viable. It may even be the case that those problems cannot be solved satisfactorily at all. In the following we will, motivated by the example above, discuss the most important problems.

Completeness problem

The most difficult question to answer is how the expert can provide a "complete" or "strong" set of property functions in advance, not knowing all the possible functions a PbD user will have in mind at any future time. "Complete" or "strong" is not formally defined; it means something like the functions must be "powerful" enough to interpret and generalize examples that were not taken into account by the person responsible for defining the property functions.

For example, the property functions of the last section are "powerful" enough to demonstrate also the concept of filling polygons with equi-distant lines or patterns. "Powerful enough" means that the properties of gradient, parallelism, distances and object relations are sufficient also for line filler functions. Fig. 13 gives an example.

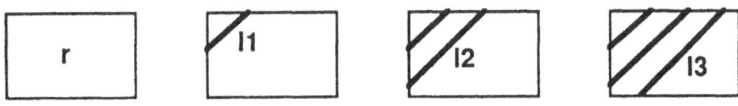

Fig. 13: Demonstrating Line Fillers for Rectangles

The instantiated function schema then looks like

$$
\mathbf{IF}
\begin{bmatrix}
\texttt{in-scene (rectangle, r)} \\
\texttt{in-scene (line, P1)} \\
\texttt{meets-2inside (P1, r)} \\
\texttt{gradient (P1, g)}
\end{bmatrix}
\mathbf{THEN}\ \texttt{drawline (P)}\ \mathbf{WITH}
\begin{bmatrix}
\texttt{gradient (P, g)} \\
\texttt{meets-2inside\ \ (P, r)} \\
\texttt{parallel-neighbor (P, P2)} \\
\texttt{distance (P, P2, d)}
\end{bmatrix}
$$

"meets-2inside" is a derived property that defines a line being inside a polygon and touching two sides of it.

World reference problem

There is a big difference between the line filler example and the 3D extension example before. The strategy for instantiating the function schema is different. In the line filler example, the precondition Pred cannot rely only on world state w_0. To find the concept of line fillers one has to reason about the world state before the current one ($= w_{i-1}$). Only that way one can find out that the distances between two neighbors are equal. The general problem, upcoming in this example, is how to determine in advance to which world state(s) pre- and postconditions have to refer. Can this decision be automated at all? Can the end-user be interactively involved to answer that question during demonstration etc.?

Sequence detection problem

Another strategy problem during the generalization step is to determine sequences of "related actions". For example, when demonstrating the 3D extension of objects, the end-

216

user will most likely extend his rectangles in any order, and not first the supporting and then the closing lines. However, that means the PbD system has to find out the difference between the two kinds of lines automatically. This leads to the well-known sequence detection or prediction problems where a program has to find out underlying rules in a sequence of numbers and predict the next one (as in the card game Eleusis, [Di86]). It remains to be seen to what extent that problem can be automated or, maybe more promising, to what extent the end-user can support the PbD system interactively.

Relevancy problem
Another problem involves the relevancy of objects during the interpretation. If the graphics screen is full of lines, rectangles and other objects, lots of relations exist and have to be calculated. To reduce the number of objects involved as early as possible, would help the PbD interpreter to keep the feature vectors small and speed up interpretation and generalization drastically. Again, the difficult question is how relevancy should be defined: is it dependent on one world, or on a few or perhaps on the whole sequence of all world states during demonstration? How far can complexity be reduced by introducing different action levels or interaction with users?

Particular problems of the 3D PbD graphics editor
The examples for 3D extension and line fillers lead at once to a lot of unsolved problems. Closing lines for concave polygons (see fig. 11) may quite easily be included by means of some extra property functions. However, property functions that are "powerful enough" to detect the concept of hidden line algorithms in general can probably not be provided without great additional effort. But hidden line capabilities will become important as soon as someone wants to draw stacks of 3D blocks correctly.
Another, much more complicated functionality is the possibility to draw 3D objects according to a central perspective view. Here, the recognition of a (virtual) vanishing point is required that may lie far outside the editors workspace.
The problem of learning non-equidistant line fillers is very similar. In that example, a time-varying distance function has to be learned. Such functions, however, can be arbitrarily complex.
Again, in both of the last two cases, the question is whether such services are learnable without providing *special* property functions in advance that already contain the concepts in some abstract form. Again the effect of user interaction has to be considered.

Further questions

Besides the problems discussed above there a a lot of questions, partly of a more practical nature. For example

1. Which structure of the world is the most useful one (e.g. alternatively to the type hierarchy in section 4)?
2. How can one embed negative examples? In which way would the generalization benefit from that?
3. How should parametrized actions be included (e.g. when searching for a certain text pattern on the screen)?
4. How far can the end-user be involved in the interpretation and generalization steps? How much access should he have to the PbD shell (see fig. 2) to define additional property functions himself? In which way can he be led to provide the "right examples"?
5. Is it possible to learn (some of) the property functions themselves (or some of the derivation rules, respectively)? This could help the "expert" to reduce the burden of programming many property functions in advance.
6. What are the theoretical boundaries of using examples to generate function code? How far can one weaken those limitations if a human is actively involved and provides "goal-directed" examples, not randomly chosen ones?

6. Conclusion

Auto-customizable software as an application of PbD is an important precondition for intelligent workstations. Software can be personalized, its development can be accelerated, and the result is often more user-oriented than those achieved by alternative methods.

However, auto-customizable software is not the only possible application of the PbD concept. It is also very useful for the generation of action sequences by demonstration (e.g. robotics or extension of diagnosis and control software) that are easy to demonstrate but difficult or time-consuming to program with conventional computer languages. Therefore, a successful application of PbD would have major economic advantages compared to the classical form of programming.

The presented architecture of the auto-customizable software supports both, tool users and developers. Tool users are provided with a graphics desk-top, usable even by computer novices. Furthermore, our concept supports the demonstration of new functionality and the guidance of the derivations when generalizing function schemata by question/answer dialogues with the user. Additionally, an explanation facility supplies advanced users with the capability of validating (or even verifying) the derivation of the generalized action sequence. Inexperienced users are provided with an UNDO function that allows them to cancel immediately the effects of an previously inferred function. Therefore, the user is able to test a new functionality simply by starting it without the danger of destroying the work done so far.

Tool developers are provided with a domain-independent interpretation mechanism (PbD interpreter), PbD demonstration system and an acquisition facility for domain-dependent information (PbD shell). With the help of these predefined modules the extension of basic tools with auto-customizable features can be carried out. Hence, our architecture of auto-customizable software tools meets nearly all the requirements of section 3.2. An expection is the requirement for a "strong" set of property functions. In general, this requirement cannot be fulfilled. In practise, the problem of a "strong" set of properties can be weakened by supplying the PbD shell to end users. They can then incrementally extend the property functions in order to demonstrate additional or more complex functions than were orginally possible.

To study the problems also from different points of view other possible applications for our PbD paradigm are considered.

1. The enhancement of OS command interpreters by useful macros. The idea of generating shell scripts by watching users typing UNIX commands, was partly discussed in [Han84].

2. The generation of new diagnosis and control functions by recording the actions of a human operator who solves a technical problem in a machining center.

3. Programming an autonomous vacuum cleaner by demonstration. Instead of putting the vehicle in an unknown environment and letting it explore from scratch, the housewife should perform "her last clean", i.e. she programs the cleaner by using it exactly once.

4. The repetition of complex assembly tasks by a manipulator. The benefit arises from the most comfortable teach-in facility one can imagine, namely if a human could just demonstrate a task with his hands and the manipulator would repeat it.

7. References

[Bau85]   F.L. Bauer et al.: "The Munich Project CIP. Vol. 1: The Wide Spectrum Language CIP-L". *Lecture Notes in Computer Science 183*, 1985, Berlin: Springer

[Car82]   J.G. Carbonell: "Learning by Analogy: Formulating and Generalizing Plans from Past Experience". *Carnegie Mellon University*, Tech. Report CMU-CS-82-126, June 1982

[Car89]   J.G. Carbonell, C.A. Knoblock, S.Minton: "PRODIGY: An integrated Architecture for Planning and Learning". *Carnegie Mellon University*, Tech. Report CMU-CS-89-189, Oct. 1989

[Cyp91]   A. Cypher: "Eager: Programming Repetitive Task by Example", *ACM CHI '91*

[DeJ86a]   G. DeJong, R. Mooney: "Explanation-Based Learning: An Alternative View". *Machine Learning*, Vol. 1, No. 2, pp.145-176, 1986

[DeJ86b]   G. DeJong: "An Approach to Learning from Observation". in: [Mic86], pp. 571-590,

[Die86]   T.Dietterich, R.Michalski: "Learning to Predict Sequences". in: [Mic86], pp. 63-106

[Han84]   S.J. Hanson, R.E. Kraut, J.M. Farber: "Interface Design and Multivariate Analysis of UNIX Command Use". *ACM Trans. on Office Information Systems*, Vol. 2, No. 1, March 1984

[Ike91]   K. Ikeuchi, T. Suehiro: "Towards an Assembly Plan from Observation: Task recognition with polyhedral objects". *Carnegie Mellon University*, Tech. Report CMU-CS-91-167, Pittsburgh (PA), Aug. 1991

[Kan91]   S.B. Kang, K. Ikeuchi: "A Framework for Recognizing Grasps". *Carnegie Mellon University, Robotics Institute*, Tech. Report CMU-RI-TR-91-24, Nov. 1991

[Kod86]   Y. Kodratoff, J.-G. Ganascia: "Improving the Generalization Step in Learning". in: [Mic86], pp. 215-244

[Kun90]   Y. Kuniyoshi et al.:"Design and Implementation of a System that Generates Assembly Programs from Visual Recognition of Human Action Sequences". in: *Proc. of the IEEE Int. Workshop on Intelligent Robots and Systems (IROS '90)*, Aug. 1990

[Mau89]   D.L. Maulsby et al.: "Metamouse: Specifying Graphical Procedures by Example". *ACM Computer Graphics, SIGGRAPH '89*, Vol. 23, No. 3, pp. 127-36, July 1989

[Mic83]   R.S. Michalski: "A Theorie and Methodology of Inductive Learning", in: *Machine Learning, An Artificial Intelligence Approach*, Morgan Kaufmann, 1983

220

[Mic86]   R.S. Michalski, J.G.Carbonell, T.M. Mitchell (eds.): *Machine Learning, An Artificial Intelligence Approach*, Vol. II, Morgan Kaufmann, 1986

[Min89]   S. Minton, J. Carbonell et al.: "Explanation-Based Learning: A Problem-Solving Perspective". *Carnegie Mellon University*, Tech. Report CMU-CS-89-103, Jan. 1989

[Mye87]   B.A. Myers: "Creating User Interfaces by Demonstration". *University of Toronto, Computer Systems Research Institute*, Tech. Report CSRI-196, May 1987

[Mye90a] B.A. Myers: "Demonstrational Interfaces: A Step Beyond Direct Manipulation". *Carnegie Mellon University*, Tech. Report CMU-CS-90-162, Aug. 1990

[Mye90b] B.A. Myers et al.: "GARNET - Comprehensive Support for Graphical, Highly Interactive User Interfaces". *IEEE Computer*, Nov. 1990

[Nis91]   F. Nishida, S. Takamatsu, Y. Fujita, T. Tani: "Semi-Automatic Construction from Speciofications Using Library Modules". *IEEE Trans. Software Engeneering*, Vol. 17, No. 9, Sept. 1991

[Ric88]   C. Rich, R. C. Waters: "The Programmer's Apprentice: A Research Overview". *COMPUTER*, Vol. 21, No. 11, pp. 10-25, Nov. 1988

[Sha81]   E. Y. Shapiro: "Inductive Inference of Theories from Facts". *Yale University, Dep. of Computer Science*, Research Report 192, 1981

[Sha89]   J. W. Shavlik: "Acquiring Recursive Concepts with Explanation-Based Learning", *IJCAI '89*, Vol. 1, pp. 688-93, 1989

[Vel89]   M.M. Veloso: "Nonlinear Problem Solving Using Intelligent Casual-Commitment". *Carnegie Mellon University*, Tech. Report CMU-CS-89-210, Dec. 1989

Groupware: A Quasi-Experimental Design for the Study of Electronic Mail

Rhea L. Walker, M. E. M., Northwestern University

Introduction

History shows that new technology changes our world. In the mid-1800s, the telegraph broke the world's geographical and temporal communication barriers. It also changed the face of business operations by connecting technology and the market place for the first time (Duboff, 1983). The railroads further influenced business operations by allowing for the mass distribution of commodities and consumer goods. The telephone first offered a technology allowing the separation of production and administration in businesses. It supported the creation of national markets in stocks and commodities (Sproull et al., 1990. The office paperwork technology (typewriter, adding machine, and new copy technology) supported the administrative office but did not change the communication process for businesses. Unlike this machinery, the introduction of computers in the early 1960s changed the number and nature of the communication process in businesses. It automated paper systems for routine tasks into computer systems and decreased the number of people required to accomplish routine transactions (Zuboff, 1988). The computer supported the further development of the international market. Computer-based technology will continue to change the reality of our world.

One important way that computer-based technology may be changing the nature of business communication is through the software known as groupware. Some researchers claim that groupware powerfully influences the interactions within and among work groups. However, little quantitative evidence exists. Such research is very important because work groups are important to the future group of the United State's industries. Organizations are flattening middle management to remain profitable. This management layer is often replaced by business teams whose development could be enhanced by groupware. Business literature, and computer and social science research claim that the use of groupware changes how groups interact to accomplish tasks. Groupware can eliminate missed communications in groups, create new decision-making methods, and

222

improve control of information for groups. It can also provide groups with new choices about their power-based relationships. Are these claims true? What changes occur when introducing groupware to a group? Is groupware a technology that changes the reality of our world?

Definition of groupware

The word "groupware" originated in business literature, but computer and social science research also define it. Dyson's (1989) definition is representative of business's attempts at defining groupware: groupware is capable of coordinating networks of computers and data for intelligent management and monitoring of work flow. Her definition falls short of being useful because it neglects to mention that groupware benefits the interaction of people in groups. Ellis et al. (1989) define groupware from the perspective of computer science research as computer systems supporting groups engaged in a common task through access to a shared environment. They limit the definition of groupware by defining it as a computer system. Groupware is a software network of computer systems. Further, Dyson and Ellis et al. do not emphasize group processes. Huber (1990), a social scientist, defines advanced information technology by proposing a theory about its use in organizations. He maintains that the availability of this technology influences its frequency of use. Frequent use creates a knowledge-based environment leading to increases in information accessibility, and this increase changes the organizational structure by improving its decision-making. If by advanced information technology Huber means groupware then his theory could predict that the use of groupware improves organization decision-making. However, his theory only provides a weak platform for empirical application because he ignores the unique nature of technology in organizations. Organizations frequently perceive the technology as a threat and therefore they do not use it (Allen, 1992). Even it the organization accepts the technology and more information becomes available, more information does not necessarily improve decision-making because it creates information overload (Alter, 1991; Young, 1992). While all of these definitions provide insight into understanding groupware, each is inadequate. A more adequate definition may be synthesized: Groupware combines software tools through a telecommunication network to enhance the work of a group engaged in the same task and having a common goal. This definition is better than those previously cited because it captures the elements that they missed, and it is useful to the business person, computer scientist, and social scientist.

Software Tools. Existing software tools are numerous (reference Appendix A), but general categories based on the geographical location and time of communication exchange between groups succinctly describe them. By updating and enhancing Johansen's (1991) analytic framework for groupware, software tools can be fitted into four types of communication: face-to-face, administrative, cross-distance, and distributed (reference Figure 1).

Face-to-face communication occurs in the same place at the same time and includes six general categories of software tools. *Electronic copyboards* are display boards that electronically transmit hand written images from the board to paper. *GDSS* are sets of software tools that facilitate face-to-face meetings and are used in specially design group decision rooms. These rooms are computer supported and designed so that each meeting member has access to a computer. The GDSS sets frequently include brainstorming, voting, and consensus forming functions to aid in the decision-making of a group. Examples of commercial GDSS include: SAMM (Software Aided Meeting Management) by Dickson, Anderson and Associates and TeamFocus by IBM (Watson, 1992). (Reference Appendix A for explanations of these tools and additional examples of existing software tools.) *Group decision rooms* are a unique type of team room because it exclusively utilizes GDSS. *Team rooms* are special meeting rooms that are supported by computer-based technology. Finally, face-to-face communication also includes the general categories of *personal computers* and *personal computer-based overhead projectors*.

Administrative communication transpires in the same place but at different times through nine categories of software tools. *Electronic messaging* creates, stores, and displays messages and then communicates them through a computer-based shared network. One group member can leave a message for another member to retrieve at a different time. Electronic messaging is also a form of cross-distance and of distributed communication and is the only category of software tools that fits into three types of communication. Commercial electronic messaging systems include: BEYONDMAIL by Beyond, Inc. (Wexler, 1992), BOZO FILTER by Agility Systems (Bennett, 1992), and EMIS (Electronic Mail Integration System) by EMIS, Corporation (Caswell, 1990).Other administrative communication categories are computer-based shared *work files, data bases, word processing*, electronic tools aiding *cross work shift communication*, computer-based tools displaying work group's *accomplishments* and *schedules*, electronic *calendars*, and computer-supported *team kiosks*.

Figure 1: A GROUPWARE TABLE

DESCRIPTION OF THE GENERAL CATEGORIES OF SOFTWAR TOOLS

	SAME TIME: FACE-TO-FACE COMMUNICATION	DIFFERENT TIMES: ADMINISTRATIVE COMMUNICATION

SAME
 	
PLACE
	Electronic copyboards
	
	Personal computer (PC)
	PC_based overhead
	  projectors
	Group Decision Support
	  Systems (GDSS)
	Group decision rooms
	Team rooms
		Shared work files, data bases,
		  and work processing
		Cross shift communication
		Group work displays, schedules,
		  and calendars
		Team kiosks or rooms with
		  computer support
		Electronic messaging

	SAME TIME: CROSS DISTANCE COMMUNICATION	DIFFERENT TIMES: DISTRIBUTED COMMUNICATION

DIF_
FERENT
 	
PLACES
	Conference calls
	
	Graphics and spread
	  sheets with audio
	  conferencing
	Video teleconferencing
	Geographically
	  dispersed GDSS
	Spontaneous
	  electronic meetings
	Electronic messaging
		Group writing and desk top
		  publishing
		Computer conferencing
		
		Voice mail
		Conversational structuring
		and distributed GDSS
		Electronic form's management
		  with signatures
		Group voice mail
		Electronic messaging

Cross distance communication is geographically dispersed communication that occurs at the same time. A commonly known software tools that provides communication across distance at the same time is the *conference call* through a telephone network. During *audio communication*, that ability to also communicate with graphics and spread sheets to the other people involved in the conversation is a less common cross-distance tool. Further, *geographically dispersed GDSS* are fairly new tools that provide communication using GDSS functions across physical distances. *Spontaneous electronic meetings* support informal "hallway gossip" communication through computer networks that provide written, verbal, or visual contact. CAUCUS by MetaSystems Design Group (Opper, 1988) is an example of such a tool that provides interactive written communication. Finally, *video teleconferencing* and *electronic messaging* are also cross distance communication categories.

Distributed communication happens anywhere at any time and represents the most comprehensive type of communication through eight general categories of software tools. *Group writing* provides a geographically dispersed work group with the ability to collaborate on writing the same document at different times. It attempts to duplicate people's activity when writing a document in a face-to-face environment. *Desk top publishing* enhances the administrative communication provided by work processing tools through support of communication at different times. *Computer conferencing* is an electronic meeting, where communication occurs through the computer-based software at different locations and times. *Conversational structuring* is *distributed GDSS* that support communication anywhere at any time. For example, an internationally dispersed work group could use GDSS functions to conduct a decision-making "meeting" attended over different times, while GDSS captures all information for exchange, documentation, and final decision-making. GroupSystems V by Ventana is such a tool (Watson). A commercial product is also being developed by Option Technologies, Inc. to support this category (Flexner, 1992). *Electronic forms' management* is computer-based communication of standardized documents. For example, a business form for travel expenses is electronically created, stored, and transmitted rather than the typical paper method. ERA (Electronic Routing System) by Hughes Aircraft Company (Caswell), NOTES by Lotus Development Corporation (Wexler), and VIEWSTAR by Viewstar (Dyson, 1990) are examples of electronic form's management. Group voice mail sends the same voice mail message to groups of people. VOXLINK by VoxLink Corporation

226

(Bennett) is such a tool. Finally, *electronic messaging* is also a category of distributed communication. The study of electronic messaging is important because it allows three types of communication: same place and different times, different place and the same time, and anywhere at any time. While it has existed for over thirty years in a limited environment, it has not been previously studied as a software tool of groupware (Sproull; Johansen). Electronic messaging demands study as a software tool of distributed communication, where it could also function as the backbone system supplying the telecommunication network.

Telecommunication network. A telecommunication network connects the software tools of groupware through shared electronic devices for access by the group. It is what Keen (1990) calls the "highway system" for the software tools of groupware. Groupware is the highway's traffic. The electronic devices are telephones, computer terminals, computers, and secondary storage devices (storage devices that save information for future processing) (Alter). Examples of telecommunication networks are electronic messaging systems, telephone systems, Local Area Networks (LANs), Wide Area Networks (WANs), and Value Added Networks (VANs). LANs usually connect personal computers and other equipment at one work location in a small work area for the benefit of electronic communication within a work group. WANs link geographically separated work locations usually through business-owned networks to also facilitate communication. For example, they could link multiple work location LANs. VANs are data networks that provide general public access to commercial data bases and software. They are usually sold to the general public for a fee and can provide them with access to general information about the stock market, merchandise, weather, etc.

Group. The group is simply a collection of individuals engaged in the same task having a common goal. The task is an analogous activity. The goal is a collective objective of the individual group members. The group can be any size in any environment. For example, a group can be one person communicating with many others about a technical problem, where everyone is responsible for the same commuter system and have a common goal to ensure its quality. Or, a person could discuss political candidates with others through an electronic bulletin board, where all discussion participants are engaged in the same task of voting with a common goal of electing the best candidate. Another example of a group that would find using groupware advantageous is a special project group, where any number of specially selected individuals are chosen to be members of the group. Group members could be from technical organizations, management ranks, different industries, and educational institutions. The identical task and mutual goal are basis criteria for defining the special project. A final example of a group is the typical decision-making group used in research about computer-aided decision-making in meetings. It is usually small, not exceeding ten individuals. A specific problem or issue demands the attention of this group (McGoff et al., 1990). The groupware task is the problem and is obviously the same one. The groupware goal is the group's need to make the best decision to solve the problem and is obviously a shared one.

Research Design

Hypotheses

This research hopes to add to the minimal quantitative study about groupware influences on group processes. Electronic mail (e-mail), a form of electronic messaging, is the software tool of groupware studied. It is administrative, cross distance, and distributed communication. In this study, e-mail is the hypothesis's independent variable or the study design treatment. Group communication, decision-making, and information exchange are the hypothesis's dependent variables. A change in group communication will be measured by comparing the number and quality of communication occurrences. Decision-making will be evaluated by measuring the time to complete decisions and judging the quality of the final decision. A change in information will be captured by comparing the accessibility and immediacy of the exchanged information. All three dependent variables will be analyzed to answer the research question: What changes occur when introducing e-mail to a group? Further, this study addresses the hypothesis: E-mail influences group processes. Hopefully, the study of this hypothesis will further the understanding of the claims made by researchers about groupware and provide more support for claiming that groupware is a powerful new technology.

228

Design

Action research combines practice with theory while ethically supporting the experimental research. Study respondents are treated respectfully as they are observed in their social setting. (Williams et al., 1988). It is especially beneficial to researchers in a experimental setting where they do not have complete control of the treatment, a quasi-experiment (Adelman, 1991). Action research for this study uses the static group comparison method to pilot the survey with data collection by personal interviewing. It uses the multiple interrupted time series method to examine the arguments about e-mail that are proposed by the hypothesis with data collection through personal interviewing. Use of an e-mail interview survey will also be investigated for use with this study. The survey is the chosen instrument for data collection throughout this research design.

Site Selection. The focus of both the pilot and e-mail studies is only one industry because prior research of groupware across industries shows that groupware's effect on groups differs as the industry changes (Walker, Spring, 1992). Action research further recommends that a site sponsor the study with considerable support to permit the researcher access over a period of time to a wide range of respondents. A research sponsor such as the Management of Information Systems (MIS) organization provides such support for this technically focused study. Additionally, the sponsor's interest in other software tools of groupware ensures their commitment to the research. Responsible people for the study from the company's work force improves researcher's access to the site's people and information. Several preliminary meetings with various respondents are necessary to determine site acceptability and whether the sponsor would benefit from the study. Because of the nature of groupware, site acceptably criteria include: 1. occupational diversity with prior access to e-mail; 2. management and non-management respondents; 3. opportunity for change to allow study of an e-mail system's affect on group processes. Further, access to respondents by researchers through e-mail would be beneficial.

The site chosen for this study met the above criteria and is referenced as "the company" in this paper's discussion. The company requested their corporate material be respected as proprietary and its identity be disguised in publication of study results. It had a history of successful research projects with educational institutions. Further, researchers had access to file, documents, and reports. The sponsoring organization was equivalent to an MIS organization which greatly aided the researcher's access to the pertinent company documents and personnel. The company's interest in other software tools of groupware was indicated by their strategic plan, which proposed enhancements and growth in nineteen additional software tools over the next five years. These tools included decision support software, shared files, and geographically dispersed graphic and spread sheet communication. The criteria requesting occupational diversity and management and non-management respondents with access to e-mail were met. The company had four individual e-mail systems: All-In-One (Digital Equipment Corporation), UNIX Office (AT&T), OfficeVision (IBM), an an internal company e-mail system. The third criterion was clearly supported by data collected in the preliminary meetings with corporate executives, where they indicated the company was demanding organizational changes to remain profitable. Further, they planned to implement a new e-mail system in the near future that would merge all existing systems into one system. This new e-mail system of choice was under study by the MIS organization. They welcomed the opportunity to obtain an academic perspective of its implementation. Business literature further showed that the company was flattening management, utilizing business teams, and obtaining groupware's tools. Finally, the researchers were also given access to the company's e-mail through All-In-One.

Pilot study design

The pilot of the survey questionnaire uses the static group comparison research method (reference Diagram 1) with data collection by non-random selection of the respondents through personal interviewing. This method involves two study groups, where one group receives treatment "X" and is then observed "O" and the other group does not receive a treatment, a control group, but is still observed "O". The treatment "X" is the new e-mail system. Observation "O" occurs through use of the survey. Even though this method has many internal and external sources of invalidity in the data collection, it is the most valid design for this type of study (Adelman). Known sources of internal invalidity are the sample selection process, mortality in the sample, interaction between the two sample selections, and possibility sample maturation. Known sources of external invalidity are the effect of selection and testing on the treated sample and possibly the study method affecting the generalizability of the data collected. The researcher must be aware of the validity issues during the analysis of the survey's effectiveness.

230

Sample selection. The population for the sample selection is personnel from the company chosen through the site selection process previously described. The sample can be fairly small for the pilot because statistical analysis of data collected from the survey is not the purpose here, but the question's effectiveness is important. Management and non-management respondents with access to an existing e-mail system match the future e-mail study's respondents. All respondents must have access to existing e-mail systems and enough respondents must have access to the new e-mail system to supply the two pilot study groups.

Diagram1:

STATIC GROUP COMPARISON RESEARCH METHOD FOR THE PILOT STUDY DESIGN

Personal interview procedures as the data gather instrument. Trained interviewers meet in pairs with each respondent for approximately two hours to pilot the survey questionnaire. Both interviewers encourage the respondent to comment on the survey's format and content. One interviewer directs the respondent to answer all questions by following the instructions provided in each section of the survey. The other interviewer scribes notes detailing the respondents answers to the questions and observation of their difficulty in understanding the questions. All respondent's questions will be answered with appropriate ethical concern by the interviewers.

The survey used in the personal interviewing for the pilot's data collection has five sections. Section 1 is an introduction to the pilot. It contains a brief description of the pilot's purpose and a confidentiality statement. The introduction remains with the respondent after review at the start of the personal interview. Section 2 collects background information on the respondents about their education, training, job assignment, and salary incentives. The interviewer asks the respondent these questions to better understand the company demographics. Section 3 focuses on the mode of communication that the respondent uses when contacting other company personnel. Respondents are asked to identify the five most frequent groups of people (or person) that they communicated with over the last month. Then, they are asked to describe the mode of communication used in that exchange (face-to-face, formal meeting, video conference, e-mail, etc.). They are further asked to identify the frequency of the communication as daily, weekly, or monthly. The respondents complete this section's communication matrix under the direction of the interviewers. Section 4 focuses on e-mail's effect on communication, decision-making, and information exchange in groups. The interviewer directs the respondents to answer a combination of open-ended and close-ended questions. The interviewer asks questions about the user satisfaction and productivity pertaining to their use of e-mail, both the existing and new systems. These questions cover topics including: reliability of access through e-mail to other people; security of e-mail messages; problem support provided by the e-mail "help" organizations; amount of use of e-mail in job assignments; accessibility of information through e-mail. Finally, Section 5 obtains information about software tools of groupware other than e-mail that are used by respondents. Respondents identify their use of tools from a list containing: electronic copyboards, voice mail, group voice mail, word processing, GDSS, data bases, computer conferencing, etc. They further identify tools useful in their job assignment that they do not presently have available to them in their work environment.

Data analysis. First, the company and its demanding work constrained the sample size, therefore sixteen of the twenty-one scheduled interviews were successfully completed to meet the pilot's requirements. Further, the sample included management (25%) and non-management (75%) respondents. All respondents accessed an existing e-mail system: All-In One (45%), UNIX Office (11%), OfficeVision (11%), and internal e-mail (33%). Thirteen percent of the total respondents accessed the new e-mail system and formed the treated study group, while the remaining 87% of the respondents formed the control group. Even though the criteria for the choice of the sample groups were met, the large difference between the treated and control group's size suggest that the results from the pilot of the new e-mail questions may be weak.

Next, interviewing was also constrained by the company demands for respondent's attention. Of the completed interviews, 44% followed the interviewing procedures, 56% modified the procedures. Modified procedures included two techniques: 1. one interviewer per respondent with increased interviewing time per respondent and 2. one interviewer with a group of respondents where a hour meeting discussed the survey's questions and then the respondents answered the questions on there own time. Despite these modifications to the interviewing procedures, the interviews provided useful information on the improvement of the survey. Interviewers further compared the data collected from all interviews and found no observable difference between the responses from the different interview procedures.

Then, the researchers evaluated the effectiveness of each survey section. The evaluation included researchers' awareness of the potentially weak analysis of the new e-mail questions and the modified interview procedures. Adding a brief description of the characteristics of the new e-mail system to Section 1 improved the introduction to the research study. Better response categories that identified consistent answers to questions improved Section 2. Combining all productivity questions into this section further improved it. Section 3 required many changes to supply the most effective data about respondent's modes for communication. These changes included collecting additional data that: identified each person's work location; typed the position of each person as a boss, peer. or subordinate; identified communication exchanges and defined them to be: personal communication, work communication, decision-making, and information exchange; supplied the duration of each message as a minute, one-half hour, or one hour. Further, the modes for communication changed to better focus on e-mail. They became: face-to-face informal conversation, face-to-face formal meeting, telephone, formal written correspondence, hand written notes, informal e-mail conversation, formal e-mail conversation rather than a formal meeting, e-mail substituted for a telephone conversation, formal e-mail correspondence, and an informal e-mail note. General updates to Section 4 improved the validity of question wording and response categories. Reformatting this section to clearly identify the purpose of each series of questions improved its readability. Since no decision-making questions were included in the survey for the pilot study, this section also added questions to better address decision-making.

Finally, minor changes to Section 5 provided updates to the suggested list of software tools. (The revised survey is located in Appendix B.)

The conclusion of the pilot study of the survey questionnaire is that while data resulted in many improvements to the survey, it did not successfully test the survey. The extensive changes to the survey coupled with the sample selection issues and the interview procedural changes invalidate the pilot of the survey. Researchers recommend another pilot study of the revised survey.

E-mail study design

After revising the survey from the additional pilot study recommended above, the e-mail study can be initiated. It uses the multiple interrupted time series research method with data collection from randomly selected samples through personal interviewing. This research method allows for the data collected from experimental groups to be repetitively compared, therefore gaining certainty about the interpretation of the causal influences of the hypothesis's argument (reference Diagram 2). (Personal interviewing may be effectively replaced or enhanced by using e-mail interviewing rather that a face-to-face exchange as the medium for asking the questions in this design after further investigation(Walker, Winter, 1992).) Random selection of the study's four groups occurs by dividing the population from the site selection process into clusters which have the common characteristic of having access to e-mail, and then randomly choosing respondents from the cluster for the study groups (Agresti et al., 1986). The personal interviews follow the same guidelines as those described in the pilot study design. Following the multiple interrupted time series research method practices, the four study groups are divided between treated and control groups. Treatment "X" will be applied to the treated groups in staggered months. Each of these groups will receive one treatment and be observed "O" before and after the treatment. The two control groups will be observed, but have not received a treatment. One control group will obviously not receive a treatment ("-"), but the second control group will be using the new e-mail system as though it functioned as just another existing e-mail system ($\underline{X}$), which is equivalent to not treatment. The duration of the e-mail study design data collection is three months and therefore observations for each of the four study groups will occur once a month for the study's duration.

Diagram2:

MULTIPLE INTERRUPTED TIME SERIES RESEARCH METHOD FOR TEH E-MAIL STUDY DESIGN

SAMPLE	MONTH 1	TREATMENT	MONTH 2	TREATMENT	MONTH 3
R	O	X	O	-	O
R	O	-	O	X	O
R	O	-	O	-	O
R	O	X̲	O	X̲	O

KEY:
R= RANDOM ASSIGNMENT OF RESPONDENTS TO GROUPS
O=A MONTHLY OBSERVATION
X=TREATMENT
"-" = NO TREATMENT
X̲= PRESENTATION OF X IRRELEVAN TO THE HYPOTHESIS

Experimental validity. Randomly selecting respondents for the four study groups by cluster sampling uses a global selection process, therefore internal validity issues of uncontrollable historical events occurring during the experimental testing interval, events of passage of time, multiple testing effects, measuring instrumentation change during the experimental testing interval, selection bias from the use of extreme pre-experimental test scores, and various selection issues (maturation, history, and instrumentation) become insignificant. The remaining internal validity concerns for the data analysis are: imitation of treatments between the treatment and control groups based on unanticipated communication; compensatory equalization of treatments; compensatory rivalry between treatment and control groups because there is a perceived less desirable application; resentful demoralization of the respondents with the less desirable application; experimental mortality. Randomizing the selection for respondents further minimizes the external validity issues surrounding the interaction between testing, selection, setting, and history, with treatment. Finally, the multiple interrupted time series design does not add back any internal or external validity issues in the data collection previously deleted as discussed above because this method is constructed to repeat observation of data. Further, its iterated data collection allows analysis by the comparison of like data over time (Campbell et al, 1970: 1979).

Survey development as the data gathering instrument. The development of the Electronic Mail Survey (reference Appendix B) for the e-mail study design used Rubenstein's (1981) organizational design paradigm (reference Diagram 3) to frame the hypothesis: E-mail influences group processes.

Diagram 3:

TABLE OF PARADIGM ELEMENTS

DESIGN FEATURES----------------O---------------->DESIGN CRITERIA
(Independent variables: | (Dependent variables:
What can we manipulate?) | What do we want to
            | accomplish?)
DESIGN PARAMETERS (CONSTRAINTS)
(What conditions must we accept as given?)

The design feature or independent variable was the new e-mail system and provided the manipulable variable for the survey (reference Diagram 4).

Diagram 4:

PARADIGM ELEMENTS FOR THE E-MAIL STUDY DESIGN

```
DESIGN FEATURES----------------O----------------->DESIGN CRITERIA
(Independent variables:         |                 (Dependent variables:
NEW E-MAIL SYSTEM               |                 GROUP
                                |                 PROCESSES:
                                |                 1. COMMUNICATION,
                                |                 2. DECISION-
                                |                 MAKING, AND
                                |                 3. INFORMATION
                                |                 EXCHANGE.
                                |
          DESIGN PARAMETERS (CONSTRAINTS)
          CONTINUOUS PLANNED REORGANIZATION
```

The design criteria or dependent variables were the group processes. Further delineation of the processes identified group communication, decision-making, and information exchange as dependent variables for realizing this study's hypothesis. Each dependent variable can then be matched with sections and questions of the survey (reference Diagram 5). The design parameters establish the conditions of the quasi-experiment that must be accepted as given constraints for the study. The main constraint for this study was the continuous planned reorganization of the company chosen as a site.

Diagram 5:

E-MAIL SURVEY DESIGN CRITERIA MATCHED WITH SURVEY SECTIONS AND QUESTIONS

E-MAIL SURVEY DESIGN CRITERIA:	SURVEY SECTIONS	QUESTIONS
1. GROUP COMMUNICATION	3	type PC and WC for all 5 groups of people
	4	1, 3-7, 12-15, 17, 19, 20, 21
	5	1 and 2
2. GROUP DECISION-MAKING	3	type DM data for all 5 groups of people
	4	22
3. GROUP INFORMATION EXCHANGE	4	8-11, 16, 23, 24

Summary

The implementation of this e-mail study design is important to the quantitative study of electronic messaging in distributed communication. This study will hopefully provide useful data in the research of the hypothesis: E-mail influences group processes, for group communication, decision-making, and information exchange.

The study is also critical to further the understanding of the distributed communication of groupware for business people, computer scientists, and social scientists and can provide additional insight into the research question: What changes occur when introducing groupware to a group? Researchers anticipate that the use of e-mail will increase the amount of group communication but decrease the overall quality of that communication. Further, it will increase the time to make group decisions but improve the quality of the decision. Finally, it will increase the quantity of information exchanged in a group but decrease the overall quality of the information.

Once successfully completed, this study design is then applicable to the research of other groupware. The treatment in the multiple interrupted time series research method could be any one of the numerous software tools identified in Appendix A. The survey questionnaire could also be modified based on the general category of the software tool under study to reflect its unique characteristics while maintaining consistent data collection of the dependent variables: group communication, decision-making, and information exchange.

Are the claims about groupware by business literature, and computer and social science research true? Does groupware eliminate missed communication in groups, create new decision-making methods, and improve control of information for groups? Can it change power-based relationships in groups? Certainly the design of this paper's research will add to the analysis of these claims to further the quantitative argument that groupware is a powerful computer-based technology that changes the reality of our world.

APPENDIX A
TABLE OF SOFTWARE TOOLS

SOME EXISTING SOFTWARE TOOL OF GROUPWARE

.

		SAME TIME: FACE-TO-FACE COMMUNICATION	DIFFERENT TIMES: ADMINISTRATIVE COMMUNICATION
SAME			AGENDA
			BEYONDMAIL
		COPE	BOZO FILTER
		OptionFinder	CIX
		PRISM	COORDINATOR
		SAGE	EMIS
		SAMM	ERA
		SYZYGY	MS-MAIL 3.0
		TeamFocus	Net Results
		VisionQuest	Office Express/grape VINE
PLACE			OFFICEWORKS
			VOXMAIL
			WordPerfect Office

	SAME TIME: CROSS DISTANCE COMMUNICATION	DIFFERENT TIMES: DISTRIBUTED COMMUNICATION
		ALEGRIA
		ASPECTS
		BEYONDMAIL
	BEYONDMAIL	BOZO FILTER
	BOZO FILTER	CIX
	CAUCUS	COORDINATOR
	COORDINATOR	EISE2
DIF-	CRUISER	EMIS
FERENT	EMIS	ERA
	ERA	ForComment
	gIBIS	GroupSystems V
	IMG	MS-MAIL 3.0
PLACES	ICICLE	NOTES
	INFORMATION LENS	Office Express/grape VINE
	SAMM	Object Lens
	TeamFocus	OptionLink
	VOXMAIL	OptionWare
		QUILT
		Rapport/Cooperator
		VIEWSTAR
		VisionQUest
		VOXMAIL
		WYSIWES

TOOL NAME: OWNER AND A BRIEF DESCRIPTION

1. AGENDA: Lotus Development Corp.
AGENDA is capable of "keeping lists of things to do and organizing meetings."
(Weixel, 1990)
2. ALEGRIA: Alegria, Corp.
ALEGRIA is a networked PC/workstation with shared data, total team/group
awareness, and real time/concurrency access. (Wagner, 1992)
3. ASPECTS: Dimitri Korahais
ASPECTS is a product that shares opinions at the same time and any location.
(Watson, 1992)
4. BEYONDMAIL: Beyond, Inc.
BEYOUNDMAIL "has foldering, filtering, and forwarding capabilities for Novell,
Inc., Action Technologies, Inc., and Message Handling System based on
networks." (Wexler, 1992).
5. BOZO FILTER: Agility Systems
The BOZO FILTER "filters junk or back burners messages " for electronic mail.
(Bennett, 1992)
6. CAUCUS: MetaSystems Design Group
CAUCUS is an on-line conferencing system that stores and forwards messages
with user interaction. (Opper, 1988)
7. CIX (Chase Information Exchange): The Chase Manhattan Bank
CIX is customized software with an easy access to information through Lotus's
Notes. (Hamilton, 1992)
8. COORDINATOR: Action Technologies, Inc.
The COORDINATOR is a PC based electronic mail system and project tracking
system that "handles correspondence among project teams and performs
commitment tracking, which entails monitoring resources." (Weixel) It can support
sharing of opinions in a different time and place orientation for communication.
(Watson)
9. COPE: University of Strathclyde
COPE assists a group in developing a shared mental model in face to face
communication. (Watson)
10. CRUISER: Bellcore

CRUISER is a video conferencing tool. (Krasner et al., 1991)

11. EIES2: NJ Institute of Technology Starr

EIES2 allows the sharing of opinions during development of lower level group's output with the capability to support any time and place communications. (Watson)

12. EMIS (Electronic Mail Integration Service): EMIS, Corp.

EMIS glues everyone on different mail systems together. (Caswell, 1990)

13. ERA (Electronic Routing System): Hughes Aircraft Co.

ERA automates paper procedures for EMIS. (Caswell)

14. ForCommen: Brodesband Software, Inc.

ForComment provides multiple review and commenting on documents. (Opper)

15. GroupSystems V: Ventana

GroupSystems is a GDSS product with multiple capabilities including: Voting, Brainstorming, etc. It supports sharing of opinions and other lower level group outputs with any time or place communications. (Nunamaker et al., 1989; Watson)

16. IMG (Issue Management Groupware): Lockheed

IMG is Lockheed's version of gIBIS (gIBIS is a graphical Issue Based Information System by MCC.) (Krasner

17. ICICLE (Intelligent Code Inspection In a C Language Environment): Bellcore

ICICLE is used for software code inspections. (Krasner)

18. INFORMATION LENS: MIT's Sloan School of Management

INFORMATION LENS is an information sharing system using artificial intelligent concepts and a graphical user interface design to create a kind of secretary that evaluates messages. (Robinson, 1991)

19. MS-MAIL 3.0: Microsoft Corp.

MS-MAIL (version 3.0) "includes a message finder feature that searches text by parameters such as subject or work." (Wexler)

20. Net Results: Lifetree Software

Net Results is a scheduling and automatic sender of electronic mail. (Opper)

21. NOTES: Lotus Development Corp.

NOTES is an "application development environment rooted in electronic mail based routing and a shared document data base used for managing corporate data flow." ("Wexler)

22. Office Express/grape VINE: Institute of IT

Office Express shares opinions within a group and supports different time with any place communications. (Watson)

23. OFFICEWORKS: Data Access Corp.

OFFICE WORKS schedules meetings and automatically sends electronic mail.

24. OptionFinder: Option Technology

OptionFinder supports lower level group outputs for face to face communication. (Watson)

25. Object Lens: MIT

Object Lens provides distributed electronic communication. (Krasner)

26. OptionLink: Option Technology

OptionLink supports both lower level group outputs and the sharing of opinions in an any time and place communication. (Watson)

27. OptionWare: Option Technology

OptionWare links keypad and keyboard groupware systems by integrating software and hardware tools for the merging of minds across time and place. (OptionWare, 1992)

28. PRISM: Tandem Corp.

PRISM is a shared mental model development tool for groups in face to face communication. (Watson)

29. QUILT: Bellcore

QUILT is hypermedia that edits documents within teams. (Krasner)

30. Rapport/Cooperator: AT&T/NCR

Rapport is multimedia groupware. (Krasner)

31. SAGE: National University of Singapore

SAGE supports lower level group output with opinion sharing in face to face communications, (Watson)

32. SAMM (Software Aided Meeting Management): Dickson, Anderson and Associates

SAMM supports lower level group output with opinion sharing at the same time for any location. (Watson)

33. SYZYGY: Information Research Corp.

SYZYGY is a PC based groupware tool. (Dyson, 1990)

34. TeamFocus: IBM

TeamFocus supports lower level group output with opinion sharing at the same time for any location. (Watson)

35. VIEWSTAR: Viewstar
VIEWSTAR regulates work content and controls work flow. (Dyson)
36. VisionNet: Leadership 2000
VisionNet supports lower level group outputs for face to face communication. (Watson)
37. VisionQuest: Collaborative Technologies
VisionQuest supports lower level group outputs with opinion sharing at any time and place. (Watson)
38. VNS (Virtual Notebook System): Baylor College of Medicine
VNS is an electronic analogy to the researcher's laboratory notebook and provides: hypermedia cross referencing; multimedia integration of text, image,audio, and video; action links integrating extreme programs; conferencing in real time; interface with graphics which are object oriented with direct user manipulation. (Burger, 1992)
39. VOXMAIL: VoxLink Corp.
VOXMAIL "converts e-mail messages to speech . . . also, it can communicate by both voice or electronics . . . by phone or computer." (Bennett)
40. WordPerfect Office: Word Perfect Corp.
WordPerfect Office schedules meetings and automatically sends electronic mail. (Opper).
41. WYSIWES (What You See Is What Everyone Sees): Xerox PARC Colab
WYSIWES allows groups to communicate by interfacing their existing tools. (Krasner)

APPENDIX B

ELECTRONIC MAIL SURVEY
SECTION 1: INTRODUCTION

Memorandum for Record   date:

To: Study Respondents

From: Interviewer

    We invite you to participate i n a research study of a new electronic mail (e-mail) system sponsored by your company's Management of Information System's organization and Northwestern University (NU). This is not a mandatory participation.

    You are part of the study of the new e-mail system, providing data about the user perception of this system and the existing systems. The study runs for three months. NU will collect the data through a personal interview with you that contains 5 sections of an e-mail survey. This introduction is Section 1 of the survey. You may keep it for future reference. The additional four sections of the survey collect data about: background information (Section 2), communication (Section 3), e-mail (Section 4), and groupware (Section 5). Collection of these data will provide NU researchers with valuable information for future study.

    The new e-mail system will provide "timely highly reliable, user friendly electronic messaging service deliverable on the standard operating environment. . . " and further "create, edit, delete, send, reply to, store, and forward electronic messages to other *company* employees on any of the existing e-mail platforms . . . of the *company*" (Request for Proposal, 1992).

    All information obtained in this study that can be identified with you will remain confidential. In any written reports or publication only aggregate data will be presented. In return, NU will provide you with a final "feedback" report, which summarizes your aggregate participation in this study. Additionally, NU requests that you do not discuss any aspects of this pilot with other personnel because such discussion may lead to distortion in data collection and analysis.

If you have any questions, please ask us during the interview. If you have any additional questions please contact us and we will be happy to answer them.

SECTION 2: BACKGROUND INFORMATION

RESPONDENT'S NUMBER: DATE:

INTERVIEWER'S NUMBER:

INTERVIEWER'S DIRECTIONS:

All questions are directed to the respondent by the interviewer. Circle the responding category to answer a question or provide a written response in the space provided.

QUESTIONS

1. What is your educational background?
Please explain.

         HS   BA   MA/MS   PHD.   OTHER

2. How many years have you worked for this company?

         0-1   1-4   5-9   10-14   15+

3. How many years have you worked in your present position?

         0-1   1-4   5-9   10-14   15+

4. What is your job category? Please explain (optional).

         management   administrative   technical

5. What is your job level? Please explain (optional).

      management   non-management

6. How many hours do you work during a typical week?

      less than 40   exactly 40   more than 40

7. How many hours do you work during a typical day?

      less than 8   exactly 8   more than 8

8. Could you briefly describe what your job is all about?

9. Could you briefly describe a normal day at your job?

10. How does your job meet the company's vision or mission statement? Please explain.

11. What kind of training have you received while at this company?

12. Who is your boss (optional)?

13. What is your boss's management style?

      analytic   driver   expressive   amiable

KEY: 1. ANALYTIC: YOUR BOSS NEEDS ALL THE FACTS BEFORE MAKING A DECISION REGARDLESS OF THE TIME INVOLVED.
2. DRIVER: YOUR BOSS NEEDS SOME FACTS TO MAKE A QUICK DECISION.

3. EXPRESSIVE: YOUR BOSS NEEDS MINIMAL FACTS TO MAKE AN
IMMEDIATE DECISION.

4. AMIABLE: YOUR BOSS WANTS TO BE YOUR FRIEND, THEREFORE
ANY STYLE OF COMMUNICATION YOU USE IS OK.

14. Who evaluates your performance (immediate boss, peers, or other)?
Please explain?

15. What are the criteria for you performance (annual goals, changing
goals, no goals, other)? Please explain.

SECTION 3: COMMUNICATION

(The following section only shows the question for 1 group. It is duplicated 4 more times to complete this section's questioning.)

RESPONDENT'S NUMBER: DATE:
INTERVIEWER'S NUMBER:
                    INTERVIEWER'S DIRECTIONS:

Tell the respondent to list the five most frequent groups of people (or
people) that they communicated with in the last month. Have them check
all appropriate modes for communication on the left and then circle the
number that best represents the duration, frequency and type of the
message on the right. Follow the key below.
                    Key:

1. To identify each person's level, please use:
    B = boss
    P = peer
    S = subordinate
    O = other.

2. To indicate the duration of the message, please use:

    5 = a minute

    4 = 15 minutes

    4 = 30 minutes

    2 = 45 minutes

    1 = 1 hour or more.

3. To indicate the frequency of the communication, please use:

    3 = daily

    2 = weekly

    1 = monthly.

4. To identify the type of communication with the co-workers, please use:

    PC = personal communication

    WC = work communication

    DM = decision-making

    IE= information exchange

    OC = other communication, please explain.

1a. Group name (optional): _____ 1b. Level: B----P----S----O

1c. Modes for communication:

  __Face-to-face informal conversation:

        duration: 5-----4-----3-----2----1

        frequency: 3----------2----------1

        type: PC-----WC-----DM-----IE-----OC

250

__ Face-to-face formal meeting:
        duration: 5-----4-----3-----2----1
        frequency: 3----------2----------1
        type: PC-----WC-----DM-----IE-----OC
__ Telephone:
        duration: 5-----4-----3-----2----1
        frequency: 3----------2----------1
        type: PC-----WC-----DM-----IE-----OC
__ Formal written correspondence:
        duration: 5-----4-----3-----2----1
        frequency: 3----------2----------1
        type: PC-----WC-----DM-----IE-----OC

__ Hand written note:
        duration: 5-----4-----3-----2----1
        frequency: 3----------2----------1
        type: PC-----WC-----DM-----IE-----OC

1d. E-mail as a mode for communication:

__ Informal e-mail conversation:
        duration: 5-----4-----3-----2----1
        frequency: 3----------2----------1
        type: PC-----WC-----DM-----IE-----OC

__ Formal e-mail conversation:
        duration: 5-----4-----3-----2----1
        frequency: 3----------2----------1
        type: PC-----WC-----DM-----IE-----OC

__ E-mail (replacing a telephone call or message):

        duration: 5-----4-----3-----2----1

        frequency: 3----------2----------1

        type: PC-----WC-----DM-----IE-----OC

__ Formal e-mail correspondence:

        duration: 5-----4-----3-----2----1

        frequency: 3----------2----------1

        type: PC-----WC-----DM-----IE-----OC

__ Informal e-mail note:

        duration: 5-----4-----3-----2----1

        frequency: 3----------2----------1

        type: PC-----WC-----DM-----IE-----OC

SECTION 4: E-MAIL

RESPONDENT'S NUMBER: DATE:

INTERVIEWER'S NUMBER:

INTERVIEWER'S DIRECTIONS:

Direct the respondent to complete the following questions. Have them circle answers when appropriate and explain in writing in the space provided when request.

GENERAL QUESTIONS:

1. What other technology besides e-mail would help you communicate better? (You may or may not already be using these technologies.) Please explain.

2. Do you use e-mail?

YES   If you answer "YES", please continue to question 3.

N O   **If you answer "NO", please skip to Section 5.**

3. What e-mail systems do you use?

_____   _____   _____   _____

4. How often do you use e-mail?

daily   weekly   monthly   other

5. Are the resources available to you to help with e-mail problems adequate? Please explain.

6. Does e-mail help you accomplish your job assignment? Please explain.

7. Please check as many of the appropriate statements which typify the relationship between e-mail and the most important functions of your job.

__ It allows me to do useful things that I could not do with out it.
__ It allows me to do the same tasks but better.
__ It makes my job easier.
__ It make my job more difficult.
__ Other. Please explain.

8. When you have a problem created by your job assignment,m does e-mail help you solve it? Please explain.

9. When you have a bright idea created through doing your job assignment, does e-mail help you communicate it to others? Please explain.

10. Are costs incurred by you in using e-mail (money, time, other resources)? Please explain.

11. Are their barriers to using e-mail (no paper trail, no response to messages, slow response, impersonal medium, junk mail, to much information, etc.)? Please explain.

EXISTING OR NEW E-MAIL SYSTEM'S QUESTIONS:
12. How convenient are the e-mail systems to use? Please check the appropriate systems and statements that you use. Please explain (optional).

existing   new
____   ____ a. I rarely use e-mail, because it is to difficult to use.
____   ____ b. I occasionally use e-mail, because it is inconvenient.
____   ____ c. I always use e-mail when appropriate, because there are
                     no disrupting barriers to my use.

13. How reliable are the e-mail systems? Please check the appropriate systems and statements to describe reliability. Please explain .

existing   new
____   ____ a. E-mail is reliable.
____   ____ b. E-mail is not reliable.

14. How secure are the e-mail systems? Please check the appropriate systems and statements to describe security. Please explain .

existing   new
____   ____ a. E-mail is secure.

15. Please check the e-mail system and statement that best describes the training you had. Please explain.

existing   new
____   ____ a. the training was a rewarding and a valuable introduction
                     to the system.
____   ____ b. Some instruction is important, but the training included
                     more that was needed.
____   ____ c. The training was less than adequate.
____   ____ d. I received no training.

16. Do any of the e-mail systems affect the way you find and receive information? Please explain.

17. Do you communicate with more people because you use an e-mail system? Please explain.

254

18. Do you use the new e-mail system?

> YES   If you answer "YES", pleas continue with question 19.

> **N O**   **If you answer "NO", please skip to Section 5.**

19. Will the use of the new e-mail system change the tasks of your job (add, delete, or change the process of the tasks, etc.)? Please explain.

20. Will the use of the new e-mail system change the way you do your present work (quality or speed of work, etc.)? Please explain.

21. Will the use of the new e-mail system change the way you communicate with others (amount or quality of communication, etc.)? Please explain.

22. Will the new e-mail system change how decision-making occurs on your job (faster or higher quality decisions, etc.)? Please explain.

23. Will the new e-mail system change how information is exchanged on your job (accessibility or speed of documentation access, etc.)? Please explain.

24. Do you get a quicker response to your problems, issues, and ideas as a result of the new e-mail system? Please explain.

25. Do you expect the answers to questions #19 through #24 to change with continued use of the new e-mail system? When? Please explain.

26. Do you expect the answers to questions #19 through #24 to change as a result of company wide installation of the new e-mail system? Please explain.

27. Do you have any other expectations of change resulting from the company wide installation of the new e-mail system? Please explain.

SECTION 5: GROUPWARE

RESPONDENT'S NUMBER: DATE:

INTERVIEWER'S NUMBER:

INTERVIEWER'S DIRECTIONS:

Give this section to the respondent to complete while you are still available, because questions may arise from the tools list. Have the respondent check any appropriate lines indicating currently using a software tool at home or at work. Then have them complete question 2.

1. Currently using software tools:

HOME	WORK	SOFTWARE TOOL:
_____	_____	a. Electronic copyboards.
_____	_____	b. Cross work shift communication..
_____	_____	c. Electronic team rooms.
_____	_____	d. Conference calls.
_____	_____	e. Voice mail.
_____	_____	f. Video teleconferencing.
_____	_____	g. Electronic forms' management
_____	_____	h. GDSS
_____	_____	i. Data bases.
_____	_____	j. Electronic bulletin boards.
_____	_____	k. Desk top publishing.
_____	_____	l. Electronic calendars.
_____	_____	m. Project management software
_____	_____	n. FAX.
_____	_____	o. Automated telephone calling.
_____	_____	p. Audio conferencing with visual graphics.
_____	_____	q. Distributed FAX.
_____	_____	r. Other:

2. What tools do you not have that you would like to have available for you use? (Please choose from the above list or add some new software tools.) Please explain.

256

Acknowledgement

Acknowledgement is extended to the following people for their efforts
in the pilot study of the survey: D. Close, C. Ellis, A. Rubenstein,
J. Sirimongkolkasem, and G. Summers. Further appreciation is
extended to J. Edy for her tireless support during the writing of this paper.

References

Adelman, L. "Experiments, quasi-experiments, and case studies: a review of
empirical methods for evaluating DSS." IEEE Transactions on
Systems, man, and cybernetics. 3/4-1991. 21:2. 293-301.

Agreski, A. & Finlay, B. 1986. Statistical methods for social sciences.
Dellen Pub. Co.

Allen, L. Personal interview by R. Walker on 5-24-92

Alter, S. Information systems - a management perspective. 1992.

Bennet, J. "E-mail moves way beyond text messages." Crains. 5-11-92.
112.

Burger, A. SIM meeting notes on 6-10-92.

Campbell, D. T. & Stanley, J. C. (1970). Experimental and quasi-
experimental designs for research. Chicago. Rand McNally & Co.

Caswell, S. "Mail messages that cut red tape." Datamation. 1-15-90. 47-50.

Cook, T. C. & Campbell, D. T. (1979). Quasi-experimentation. Chicago.
Rand NcNally College Pub. Co.

Dyson, E. "Directing computer traffic." Across the board. 1/2-89. 52-53.

Dyson, E. "Why groupware is gaining ground." Datamation. 3-90. 52-6.

Duboff, R. "The telegraph and the structure of markets in the United States, 1845-1890." Research in economic history. 1983. J. A. Press, Inc. 8. 253-77.

Ellis, C., Gibbs, S. & Rein, G. "Groupware: some issues and experiences." Communication of the ACM. 1-91. 34:1. 38-52.

Flexner, B. SIM meeting notes on 6-10-92.

Hamilton, R. "Chase banks on "info" access." Computerworld. 1-92. 6.

Huber, G. "A theory on the effects of advanced information technologies on organizational design, intelligence, and decision-making." Organizationa and communicatin technology. Fulk, J. & Steinfield,C. (eds.). 1990. Sage Pub. Newbury Pk.

Johansen, R. "Groupware: future directions and wild cards." J of organizational computing. 1991. 2:1. 219-27.

Keen, P. "Telecommunications and organizational choice." Organizations and communication technology. Fulk, J. & Steinfield, C., eds. 1990. Sage Pub., Inc. Newbury Pk, CA.

Krasner, H. McInroy, J. & Walz, D. "Groupware research and technology issues with application to software process management." IEEE transactions on systems, man and cybernetics. 7/8-91. 21:4. 204-12.

McGoff, C. Hunt, A, Vogel, D. & Nunamaker, J. "IBM's experiences with group systems." Interfaces. 11/12-90. 20:6. 39-52.

Nunamaker, J., Vogel, D., Heminger, A., Martz, B. Grokowski, R. & McGoff, C. "Experiences at IBM with group support systems: a field study.: Decision support systems. 1989. 5. 183-96.

Opper, D. "Making the right moves with groupware." Personal computing. 12-88. 135-40.

Optionware. 1992. Option Technologies, Inc. Mendota Heights, MN.

258

Robinson, M. "Through a lens smartly." <u>Byte</u>. 5-91. 177-80.

Rubenstein, A. "An organizational design paradign." . conference paper.
    1992

Sproull, L. & Kresler, S. 1990. <u>Connections</u>. The MIT Press: Cambaridge.
    Mass.

Wagner, G. SIM meeting notes on 6-10-92.

Walker, R. "A study method of the electronic-mail and personal interview
    modes of data collection: similarity of data collection and accessibility
    of respondents." working paper. Winter, 1992.

Walker, R. "Groupware issues for management." working paper. Spring,
    1992.

Watson, R. (editor). <u>Groupware report</u>. 1992. Athens, GA.

Weixel, S. "Flat managaement requires juggling." <u>Computerworld</u>. 4-27-92.
    70-1.

Wexler, J. "Filtering the message deluge." <u>Computerworld</u>. 4-27-92. 37.

Williams, D. "New technologies for coordinating work." <u>Datamation</u>. 5-90.
    92-6.

Young, T. Personal interview by R. Walker on 5-24-92.

Zuboff, D. 1988. <u>In the age of the smart machine</u>. Basic books, Inc.: NY.

Knowledge-based Telecooperation

Astrid Scheller-Houy (Siemens AG)

Ruth Bartels (DFKI)   Dieter Scheidhauer (Siemens AG)

Abstract

Telecooperation is a part of the research field Computer Supported Cooperative Work (CSCW). Geographically and temporally distributed agents (people, systems) are trying to solve a problem in a team by using a suitable infrastructure. Telecooperation systems systems support this kind of work. But using telecooperation systems means the formation of new problems.

Because telecooperation systems are very complex, their design and implementation require a high measure of experiences and knowledge. But such a comprehensive knowledge will be available only after many years of professional experiences. Often the same problems occur in the design of different telecooperation systems, because they possess same or similar elements, relations and functionalities. Nevertheless they are analysed, defined and implemented at every development of a new telecooperation system. This leads us to the idea of designing a development environment for "the construction of cooperative applications". Therefore our aim is the realization of a Knowledge-based Multimedia Communication Shell (KMC-Shell). It will be simplifying the development of telecooperation systems. In keeping with the user requirements such systems can be development with the least possible programming effort. The 4-Level-Model of Telecooperation forms the basis of the KMC-Shell. It reflects the process of refinement, which must be performed in order to develop a telecooperation system for a special application.

Introduction

Changes in the world of work cause an increasing decentralisation as well as a dense interweaving of national and international business relationships. Today a modern enterprise can no longer survive as an isolated unit, it must be thought in terms of a networked organisation. This necessitates that people who are geographically distributed have to contact each other and have to work together inspite of long distances. Apart from this, the tasks are getting so demanding and complex that it is impossible for one person alone to accomplish them. Thus a team is needed.

A group is able to come to a completely different result as any of the same people would have come to when working alone. Several people work together on one task whereby their different knowledge and different capabilities complement one another. The efficiency of a group with well functioning communication is much higher than the sum of the abilities of each individual team member. This entails that the traditional lone warrior gets also increasing needs for cooperation, discussion, and exchange of information in group situations.

In addition to this, important decisions that are often to be taken fast require an effective cooperation between organizations and people over long distances. This shows that business and administration of the future will rely more than ever on the work on teams. At this point well developed problem solving solutions, existing commercial communication tools, like telephone, telefax, e-mail and tools designed for single-user like an editor reached their limits. They can not ensure a sufficient support for an effective distributed cooperation, as mentioned above. Therefore, a new generation of systems and technologies with completely new performance characteristics is necessary, which are especially conceived for computer supported cooperative work.

Telecooperation

The research field dealing with computer supported cooperative work is given the acronym CSCW. This research field comprises investigations of how people work together in groups and how computers and related technologies have impacts on group behaviour, on an improvement of the effectivity of a group and on the work itself.

It is not easy perhaps impossible to find an overall definition of CSCW. There are many other terms connected with this abbreviation. Some of them are 'Groupware', 'Group Decision Support System', or 'Electronic Meeting Systems' to mention only a few.

The term 'Groupware' is sometimes used synonymously with CSCW [Naff90], [Non88], but it is also used for commercial CSCW tools or products in particular [Gre91], [EGR91]. A good overview on existing groupware systems is given by Hayes [Hay92].

Group Decision Support Systems (GDSS) are computer based systems supporting the solving process of unstructured or partially structured problems by decision makers who are working together as a team [DSG85], [Krc89].

Electronic Meeting Rooms (EMS) apply information technology to support meetings, where geographically distributed and temporarily shifted teamwork is possible. EMS is considered as a merge of GDSS and CSCW. GDSS is task-oriented designed and CSCW communication-oriented [DGJNV88].

Telecooperation is the part of CSCW that deals with the cooperative work of geographic distributed people and systems. This research field emphazises, as it's name indicates, remote cooperation. It analyses computer supported work covering long distances. Geographically and temporal distributed people and systems are trying to solve problems in teamwork by using a suitable infrastructure.

First systems for telecooperation are being developed but in most cases they are still at an early stage and exist only as prototypes [BJLSV92]. Although different telecooperation systems support different applications, these systems have many common features. If for instance we compare a telecooperation system for distant learning with one for ordering, both systems must support joint working on a common document (joint editing). Distant learning enables a trainer to teach a learner, who is located far away from his place. Sometimes the trainer has to make his remarks on the document, on which the learner is working at. This case happens, when the learner has problems doing his exercises on his own. Then the teacher can support the learner putting remarks in the learner's document using joint editing. The other example for the necessity of joint editing is the mentioned telecooperation system for ordering. Customer and salesperson are separated from each other. The customer wants to order some goods, therefore he has to fill out an order form. But in difficult cases for example ordering complicated machines, the customer needs help from the salesperson to order the things he wants to. Then both of them must have the opportunity to see and work on the order form. Joint editing is necessary in order to make sure that the salesperson and the customer fill in the same document that includes an order form.

Requirements of Telecooperation Systems

Telecooperation systems must fulfil some special qualifications in order to enable a cooperation between geographically distributed persons and systems.
On the basis of our experiences we expect from a telecooperation system following minimal requirements:
- joint working on documents
  (joint editing, joint drawing)
- facility of an audio conference
  (audio conference)
- access to suitable resources
  (resources managing)
- sending and receiving of electronic mail
  (group bulletin board).

Only systems integrating and supporting such tools and services are called telecooperation systems. This means all minimal requirements have to be met.
Joint working on documents enables team members synchronously exchanging of information as well as an audio conference enables synchronous information exchange. In this manner the team members can discuss and jointly edit a document. This ensures spontaneity and direct information exchange, which is required to an effective cooperation. The access to suitable resources is necessary so that the team members can use this aid for solving the common problem. But since a cooperation consists not only of meeting activities but also of no meeting activities, it is important to have the possibility of sending messages to every involved person at every time.
According to this definition, audio / video conference systems or intelligent mail systems on its own are no telecooperation systems, they are only a part of such systems. Additional prospective features of a telecooperation are joint pointing or the access to a lifeboard, for instance.
The quality of such systems depends on:
1.   the available infrastructure and
2.   the efficient and flexible usage of this infrastructure in the sense of cooperation.
Infrastructure represents workstations, devices, networks, services and so an. Applying tools or services depends on networking and device configurations. For instance, performing an audio / video desktop conference requires special device configuration and networking.

Using telecooperation systems there are arising new problems. They concern users as well as developers of telecooperation systems. Telecooperation makes it possible for a group of separated participants to solve jointly problems, while at the same time using communication and computing resources. In order to support this kind of work a number of various tools and applications like electronic mail, bulletin boards, co-authoring systems, joint editors, group schedulers, teleconferencing tools has been developed and work is still continued. Cooperative work takes place in the framework of a computer supported conference and generally involves several kinds of tools/applications at the same time. Thus, the increasing toolset demands the whole attention of the user and easily exceeds his capabilities.

The overcharge of the users must be avoided to assure the acceptance of telecooperation. Since this requirement has to be considered as early as possible, it is already a strong condition for the design process of a complex telecooperation system. Researchers and developers are required to have a great amount of system knowledge and also detailed knowledge about the application domain. Such a comprehensive knowledge will be available after many years of professional experiences. This entails that adequate personal will be seldom available. On the other hand, current projects in which our research group is involved shows us, that a greater number of generic tools could be extracted, especially coordination and cooperation mechanisms.

The ideas depicted above lead to a design of a development environment for "the construction of cooperative applications". Therefore our research objective is the realization of a knowledge-based multimedia communication shell (KMC-Shell).

KMC-Shell

The Knowledge-based Multimedia Communication Shell (KMC-Shell) is planned as a development platform which allows the construction of cooperative applications in a distributed, heterogeneous environment supported by multimedia tools. It will be simplifying the development of telecooperation systems. In keeping with the user requirements such systems can be developed with the least possible programming effort. Telecooperation systems are very complex systems. Their design and their implementation require a high measure of experiences and knowledge, as mentioned above.

Often the same problems occur in the design of different telecooperation systems because they possess same or similar elements, relations and functionalities. But they are analysed, defined and implemented at every development of a new telecooperation system.

Supposing, the developer has the order to develop a telecooperation system for a special application. At first he has to get to know the domain of the application in order to concipate the application. Therefore, he ascertains
1. the cooperation subject
2. the cooperation process
3. the involved persons and
4. the necessary means.

The cooperation subject represents the matter of the choosen application, 'distant learning of the desktop program FRAMEMAKER', 'cooperative purchase order processing', 'trouble shooting' for instance. The cooperation process reflects the executing of the cooperation process. Self learning or team learning are different ways to perform the 'distant learning process', for instance. Persons and means which are necessary for the special cooperation are deduced from the cooperation process. A feasible cooperation process depends on available persons and means. Means comprises devices and networking (LAN, WAN and so on) as well as material for the cooperation process (lexicon, videofilm) and for the communication (joint editing und so on). Using means ensure acceptable support of the distributed interpersonnel communication with regards to the contents as well as technical. Let us consider the domain 'distant learning' in order to illustrate the notions above. Developing telecooperation system for 'distant learning for FRAMEMAKER' (the cooperation subject) requires information of the cooperation process 'distant learning' and the involved person and means. For example following questions must be answered:

- Who is involved? - Trainers and learners
- How many trainers are involved?
- How many learners are involved?
- What kinds of learning methods should be supported? - Self learning, team learning and so on
- What kinds of learning material is necessary for what kind of learning?
- What kind of learning material is available?
- Should the system meet the high requirements of virtual classrooms?
- What kinds of cooperative tools are necessary?
- Is some infrastructure already existing?

In the case of 'yes'
- How is the teacher or are the teachers equipped with?
- How are the learners equipped with?
- How are the teacher and learners connected?
- What kinds of cooperative tools are available?

This is only an extract of the questions the devoloper has to answer during the developing process. This shows, that the developer must be endowed with programming attainments in order to adapt the tools to the domain and to the needs of the users, for example. He must know all about new technologies in order to find out a suitable infrastructure and he must also have knowledge about the domain respectively the application itself. It is nearly impossible to find such a person. Although the domain 'distant learning' allows developers intiutively infering on the basis of common sense and of his own experiences, it is difficult to develop a telecooperation system for this domain. But it is much more difficult within a domain like 'distant troubleshooting in the aircraft area'. Consider only the simple question 'Who are the involved persons....? (Technician, supervisor, troubleshouter, production engineer and so on) ....and what are their roles?' All this, an accumulation of various knowledge in one person and the same problems deriving by the designing of different systems, is strongly suggestive of the motivation developing expert system shells. The first expert system shell were caused by extracting all infomation from an existing expert system. These informations concerned the special application, where it was written for. The rest is called shell of an expert system.

The same proceedings is not possible for the KMC-Shell, because there are no existing telecooperation system of a special application, which is suited. Therefore, the KMC-Shell must be concipated to be easy to extend and to modify.

4-Level-Model of Telecooperation

We designed a model, called the 4-Level-Model of telecooperation which builds the common basis for different telecooperation systems. With the aid of this model the similarities of the different telecooperation systems are to be discovered with the objectives to simplify those systems and to avoid superfluous development work. The 4-level-model of telecooperation contains the various abstraction levels of knowledge which are necessary for cooperative distributed working. The levels are based on one another whereby the information of a higher level is made concrete on the one below. The top level contains the knowledge frame for the remaining levels. This knowledge frame is filled with detailed information covering the lower levels until suitable application is received on the lowest level. The lowest level can to some extend be considered as the concrete feature of the highest level.

For the development of the 4-level-model, the system theory offers its services. It is used in the framework of organization theory as a general method of investigations. A system is defined as a complex of elements with relationships between them. This point of view allows a selective accomplishment of complexity by making it possible to concentrate on the essentials of a problem. Due to the splitting up into part systems the complexity can be reduced and processed [Rem89]. A system is registered and demarcated according to the respective research reason. In the framework of telecooperation, a system can then be understood as a cooperative application. At a system theoretical consideration the elements are not of interest in their respective totality, i. e. with all their features and ways of behaviour also outside of the examined system but exclusively according to system related ways of behaviour [Gro78]. Consequently at the development of the 4-level-model, the elements are considered only in their function as performance carriers according to cooperative applications.

The 4-level-model of telecooperation (s. figure) consists of:
1.   the General level
2.   the Domain level
3.   the Application level
4.   the Runtime level

These four levels reflect the process of refinement, which must be performed in order to develop a telecooperation system for a special application.

The General level
This level comprises the necessary objects for the exchange of information during a cooperation process, their relationships and their respective attributes. The description is given in such an abstract way that it can be used for each telecooperation system. The description will become more detailed with descending levels. The general level provides - as it's name already indicates - general knowledge about telecooperation systems. This general knowledge comprises, for example, the knowledge that a cooperative process consists of persons, machine systems, computers, peripheral devices or services and that organizational and cooperation relationships within the team have to be taken into account.

If one looks at the frame of a person at this level, it can be stated domain-independently e.g., that the person has to be identifiable, that he has rights and duties which depend on the organizational and cooperative relationships, that he has access rights to resources, that he has a toolset of communication facilities and how he is accessible.

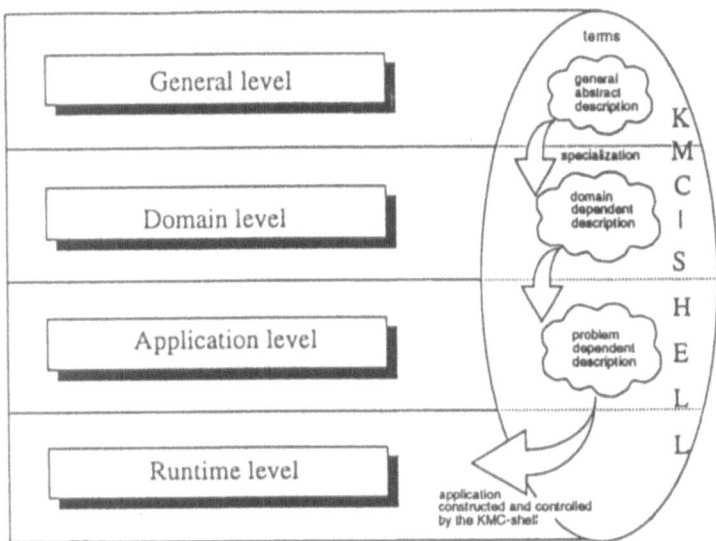

Figure: The 4-Level-Model of Telecooperation

These frames will be filled with more and more detailed information, until on the runtime level, a concrete real human is described. On that level, the human's rights and duties within the current problem solving process are known as well as his different network connections, the name of his end device etc.

Domain level

If we consider a particular application domain like cooperative distant learning or cooperative purchase order processing the question arise: what are the requirements of a telecooperation system in these domains? On this second level we characterize now more exactly the objects according to its application domain. Therefore, we identify the objects, operations, and relationships that domain experts perceive to be important about the domain. The objective is to describe objects and features which are domain specific with those from the general level. In the domain "education", for example, there is one trainer, several learners, and some teaching material. The trainer and the learners are persons, the teaching material could be a lexicon or a video film. Based on these descriptions, we are able to infer from the knowledge of the general level about the relationships of the objects, one or more structures of a telecooperation system, which will be suitable for this application domain. For example, we have a video film as teaching material and a learner wants to have access to it. His desire makes only sense, if he has the necessary hardware equipment and powerful enough network connections. Because of his role as a learner, he has only read access to this resource.

Application level

This third level provides a real team cooperation environment for a real execution of a problem solving process described on the application domain of the above level. On the problem specific level we know which persons sitting at which workstations, we know about the connections between them, about the facilities of the workstations. Provided by this information, the system consults the user with an according description of the problem. In order to step from this level to the runtime level, described below, a concrete telecooperation system has to be build. Connections are enabled, resources are brought in and services made available.

Runtime level

The runtime level represents the implementation of a special cooperative application, where the requirements, postulated in the three levels above, should be taken into consideration. Modifications, which occur during in the runtime phase, like changing team members or technical equipment, are handled by the available knowledge. If the modifications will have effect on the current structure, it could be changed or a new structure could be generated. With the KMC-Shell we are able to construct and control a telecooperation system, for example for the application domain cooperative purchase order processing.

To collect experiences as soon as possible about working in a distributed cooperative environment we have realized a working environment, where it is possible to habe audio/video conferencing, synchronous joint editing of documents and the current access to resources. If you want more information about this working environment and about the domain "distant learn" you should read the artikel of Jean Schweitzer [Sch92] in this document.

References

[BJLSV92]   R. Bartels, A. Jarczyk, P. Löffler, A. Scheller-Houy, G. Völksen: "Computer Supported Cooperative Work (CSCW) - State of the Art", internal paper of DFKI/Siemens, München-Saarbrücken 1992.

[DGJNV88]   A.R. Dennis, J.F. George, L.M. Jessupp, J.F. Nunamaker Jr., D.R. Vogel: "Information Technology to Support Electronic Meetings, MIS Quarterly, vol. 16, 1988.

[DSG85]   G. DeSanctis, R.B. Gallupe: "Group Decision Support Systems - A new Frontier", DATA BASE, vol. 16, no. 2, 1985, 3 - 10.

[EGR91]   C.A. Ellis, S.J. Gibbs, G.L. Rein: "Groupware: Some issues and experiences", in: Communication of the ACM, vol. 34, no. 1, 1991, 38 - 58.

[Gre91]   S. Greenberg: "Computer Supported Cooperative Work and Groupware: An Introduction to the Special Edition", International Journal of Man Machine Studies, vol. 34, no. 2, 1991, 133 - 143.

[Gro91]   E. Grochla: "Einführung in die Organisationstheorie", Porschel-Verlag, Stuttgart (1978) 10.

[Hay92]   F. Hayes: "The Groupware Dilemma", Unix World, February 1992, 46 -50.

[Krc89]   H. Krcmar: "Considerations for a Framework for a CATeam Research", Proceeding of the First European Conference on Computer Supported Cooperative Work (EC-CSCW), London (1989) 421 - 435.

[Naff90]   N. Naffah: "Multimedia Applications", Computer Communications, vol. 13, no. 4, 1990.

[Non88]   Noname: "Depth Groupware", Byte, 13, no. 4, 1988.

[Rem89]   A. Remer: "Organisationslehre", Berlin-New York (1989) 183 - 184.

[Sch92]   J.,E. Schweitzer: "Telecooperation: Distributed Cooperative Work of Tomorrow"; in this document.

Adaptive User Interfaces

Uwe Malinowski, Thomas Kühme,
Matthias Schneider-Hufschmidt, Hartmut Dieterich

Siemens Corporate Research and Development
Systems Technologies
System Ergonomics and Interaction
ZFE ST SN 7
Otto-Hahn-Ring 6
D-W8000 München 83
Federal Republic of Germany
malinow@zfe.siemens.de

Abstract

This paper presents a short survey of the field of Adaptive User Interfaces (AUIs). Different approaches to a definition or at least a description of AUIs are given. There are a lot of different aspects relevant to the adaptation process. The most important aspects (goal of adaptation, stages and agents, scope of adaptation, considered information) are discussed. Finally, the approach of Computer-Aided Adaptation and its basic ideas are presented.

1 Introduction - Clarification of Terms

No widely accepted definition of the terms AUI and II has been established so far. A global description is introduced by Totterdell and Rautenbach [TR90]:

> ... the user interface can be thought of as something which sits between system and user and adapts the system to the user.

Benyon proposes a definition of an adaptive system which is more tangible but again does not cover the huge number of aspects relevant to adaptivity in user interfaces [BMJ90]:

> An Adaptive System may be defined as a knowledge-based system which automatically alters aspects of the system functionality and interface in order to accomodate the differing preferences and requirements of individual system users.

On the one hand, this definition describes a broad range of systems, as it covers changes of the interface and changes of system functionality. On the other hand, the information considered in the adaptation process is restricted to user preferences and requirements. Requirements given by the application, ergonomics or the interface itself are neglected.

Figure 1 explains an additional description of the terms mentioned above expanding a presentation proposed by Elkerton and Williges [EW89]. Early user interfaces (UI) were static. The system designer built the interface and the users had to learn how to use it. Today a more flexible interface is accepted to be state of the art. The opportunity to accommodate the interface to their own preferences is given to the users. The flexibility is usually restricted to simple changes, e.g., the change of colors, size or position of windows.

An Adaptive User Interface (AUI) supports the users in the adaptation of the interface to their own needs and preferences or performs the adaptation automatically. The focus of adaptation extends to a broader range than in a flexible interface, including functionality and the demands of the application.

An Intelligent Interface is the integration of an AUI with an Intelligent Help System (IHS), which makes context-sensitive and active help available ([Sch89], [WAC84]), and an Intelligent Tutoring System (ITS) ([SB82]), which supports the users in practising the use of the system. Our inspection of AUIs includes Intelligent Interfaces since adaptivity is a main characteristic of their components (IHS and ITS). It has to be stated that the functions provided by an IHS are often regarded as an integral part of an AUI.

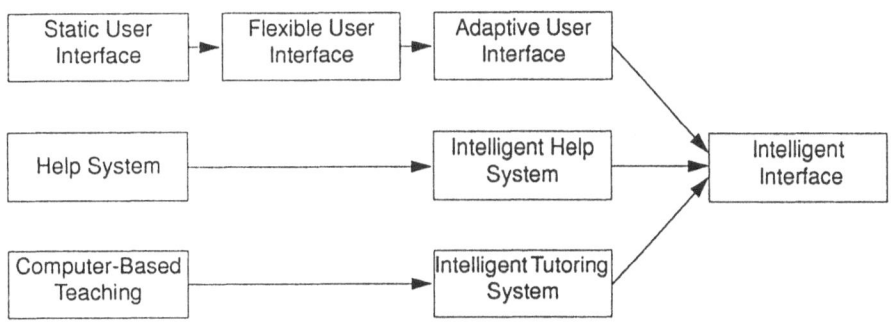

Figure 1. Adaptive User Interface vs. Intelligent Interface

2 Aspects of Adaptivity in User Interfaces

By examining about 160 papers concerning the field of AUIs, the most important aspects for classifying and describing AUIs have been identified. On the one hand, the aspects can be regarded as a taxonomy of AUI's. On the other hand, they they are to be considered in the development of an AUI [BNR90].

2.1 Goal of Adaptation

Before starting the design of an AUI the goals of adaptation have to be identified. They can be used to decide if the development of an AUI is justifiable at all. If the identified goals can be fulfilled with a flexible or even a static UI the effort necessary for the development of an AUI can be economized. Later on, during the usage of an AUI, the goals should be used as a metric for an evaluation of the AUI.

One reason for building AUIs is the heterogeneity of the group of system users. Therefore the overall goal is to make the system usable for each individual user or, more ambitious, to improve usability for the individual user.

Users of complex software systems differ very much in their level of experience. The differences appear both in the knowledge about the application domain and in the experience in using the system. Therefore a reasonable goal for an AUI is to adapt to the level of experience of the individual user.

These goals can be refined. The adaptation of a UI can aim at speeding up or simplifying the usage of an application system. Another possible goal of an AUI is to present the interface the users want or expect to see. This is not necessarily contradictory to the goals mentioned above, in many cases it results in a reformulation of the same goal.

2.2 Stages and Agents

In every adaptation process different tasks have to be performed. They can be grouped into stages of the adaptation process. While Totterdell and Rautenbach distinguish between the stages *variation*, *selection* and *testing* [TR90], we examine similar stages from the users' point of view.

The first stage, called *initiative*, is the decision of one of the agents to suggest an adaptation. Subsequently, alternatives for adaptation have to be proposed (*proposal*). In the following stage one of the alternatives has to be chosen (*decision*) and finally performed (*execution*).

Possible agents performing or controlling these stages are the system designer, the system administrator, a local expert, the user, or the system itself. If the system designer, the system administrator or a local expert performs the tasks of one of the stages, an adaptation can only consider the needs of user groups. Hence the resulting interface probably will not suit the needs of an individual user. Furthermore, in most cases it is not important to the user to detect whether an adaptation is performed by the system or the system designer, the system administrator or a local expert. Therefore, from the user's point of view the most interesting agents in the context of adaptive systems are the *system* and the *user*.

Consequently, 16 combinations (2 agents with 4 stages) have to be considered. This can be visualized in a matrix (see Figure 2). Any combination can be illustrated by marking which agent performs which tasks.

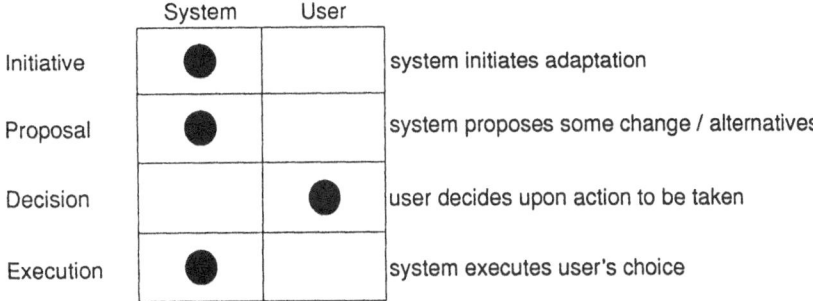

Figure 2. Tasks and Agents: Example Configuration

Among these combinations four uninteresting combinations can be identified: independently of the agent performing initiative and proposal, it is not reasonable to make the users execute an adaptation selected by the system.

If the system performs the tasks of all the stages, we call the process *Self-Adaptation*. The system observes the communication, decides when to adapt, generates and evaluates different variants, and finally selects and executes one of them. Examples of self-adaptive systems are GUSIB (Generic User Interface Builder) [Dan88], active context-sensitive help [Sch89], or error-correcting systems like FLEX [Mot90].

Self-Adaptation is particularly sensible to adapt to the requirements of the application, e.g. in ADBS [Gri85]. In contrast, adaptation to the users' needs should give control to the users. *User-Controlled Self-Adaptation* is used in POISE [Cro84] and PODIUM [She90].

Systems using Self-Adaptation or User-Controlled Self-Adaptation should allow the users to take the initiative. Those variants are designated as *User-Initiated Self-Adaptation* and *Computer-Aided Adaptation (CAA)*, respectively. Context-sensitive help, e.g. in UIDE [FWK91], is a kind of user-initiated self-adaptation. A system offering CAA takes on the routine tasks (Proposal and Execution) and entrusts the creative tasks (Initiative and Decision) to the users.

Asking the users to execute an adaptation proposed by the system is a less reasonable variant. In this case the system should also be able to execute the proposed adaptation.

In literature, there is no example of a system which asks the users to propose alternatives and performs the other stages automatically. One can imagine a system being able to evaluate proposals but unable to create them. It seems to be more reasonable to present the evaluation of the proposals to the users and by that means support their decisions.

With *System-Initiated Adaptation* the users are informed if and when it seems to be reasonable to tailor the system. Simple *Adaptation* gives the opportunity to the users to tailor a system to their own needs and preferences. E.g., almost every window manager allows the users to change colors, sizes of windows or the appearance of menus.

With respect to the knowledge necessary for a system to control each of the stages, a two-dimensional classification scheme can be introduced. We call the dimensions *System Competence for Context Analysis and Plan Recognition* and *System Competence for Proposal Creation and Evaluation* (Figure 3). The most interesting combinations of agents and tasks are placed in this scheme according to the previous text.

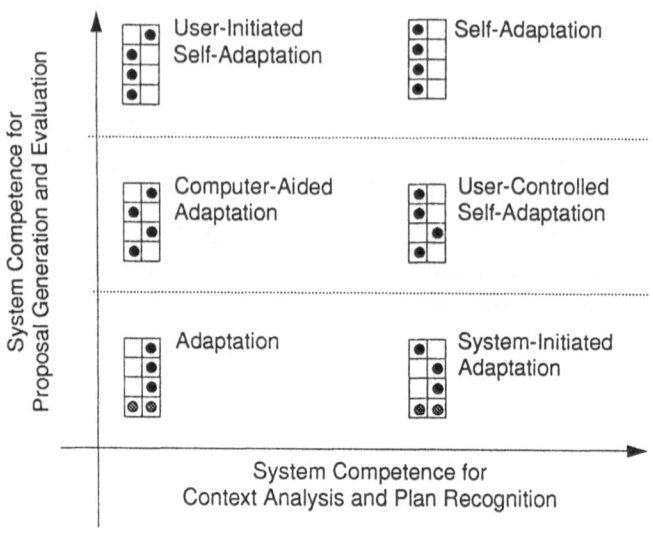

Figure 3. Classification Scheme: Tasks and Agents

2.3 Type of Adaptation

Browne states that there are no principal restrictions in the possible dimensions of adaptation [BNR90]. From a global point of view, two groups of adaptation can be distinguished. In a system using *adaptation of the communication* the users have to perform the same tasks, no matter whether adaptation takes place or not. *Adaptation of functionality* gives the users the opportunity to apply new or more complex functions.

[Cro84] and [Mot90] present systems with the ability to *correct errors and inaccurate input*. [Cro84], [Dan88], [Sch89] and other authors describe *active help systems*. These systems try to detect errors and non-optimal plans of the users during interaction with the users. On that basis they present the most appropriate information attempting to enable the users to overcome these shortcomings on their own.

Many variants of adaptation consider *type and contents of presentation*. The type of presentation is often changed by switching between several interaction styles. [BT88] describes a system with different interaction styles for experts, intermediate users and novices. Another system allows to switch between "Query and Answer", "Menu Selection" and "Command Language" ([FMS87]).

A system for process control described in [Gri85] enables the users to select every single information item which has to be presented and to change its presentation style for each information. In a critical situation occuring in the application process the user interface can decide to present information which the users normally do not want to see.

[TT89] describe a context-sensitive and adaptive user interface for UNIX. The most interesting commands are presented to the users. Subsequently the users are asked for the necessary parameters. Furthermore, plans and strategies are presented if it is appropriate for the user in the current situation.

[Mas86] presents another user interface for UNIX. The possibly required commands and accompanying information, tailored to the experience of the actual user, are presented.

A user interface for an electronic mail system presents the presumably desired parameters of a selected function to the user, dependent on the current situation and former actions of the user [BSN86]. [FWK91] describe the user interface development system UIDE. A user interface developed using UIDE provides the user with context-sensitive help.

Adaptation of functionality is even more complicated. It seems to be reasonable to allow the users to perform the creative tasks and to delegate the execution of the routine tasks to the system. [MRW85] describes a system where tasks are dynamically allocated to either the system or the user, depending on the stress of the user. The system presented in [Rou88] can perform complete tasks, propose solutions or leave the complete execution to the user.

Functional enhancement is proposed by [Ris84] for the automation of routine tasks for the individual user. This can be done by macro generation. A more elegant way is an automatic macro generation with learning by example using a generalization algorithm ([HP91], [Cot90]).

2.4 Considered Information

This paragraph presents different aspects that have to be considered in the adaptation process. Every single adaptation has to take into account all these aspects, probably with differing importance. The aspects are related to either the user, the application, ergonomic principles, or the user interface itself.

Most prototypes and systems described in the literature concentrate on the inspection of user characteristics. The most common way to consider user characteristics is the design of a system as the designer has to consider the needs and abilities of the typical user. This simple form can be supported by prototyping.

Other systems can be tailored to the needs of a user group by the system administrator. This customization can be done by programming or by switching between different system views predefined by the system designer.

In order to achieve an individual user interface it is necessary to consider information about the individual user. The individual UI can be formed by the designer, the user or the system itself. There is a large variety of different information that have been considered in prototypes. Often used terms are: needs, preferences, characteristics, abilities, interests, behavior, knowledge, experience (see e.g., [BT88], [FMS87], [BIM86], [VB88]). It has to be taken into account that there are differences between individual users and that the knowledge and preferences of the individual user changes, while using the system.

[FMS87] reduce the user's needs to the need for a particular interaction style which is predicted on the basis of the user's experience. The term "abilities of the user" summarizes many different aspects, e.g. basic abilities like using the mouse, or very complex abilities like performing a specific task in the actual application context. The latter also can be classified as a part of the user's domain knowledge.

If the system tries to correct typical errors of the user, this can be regarded as adaptation to the individual user's typical errors. [Gri85] describes ADBS, a system for process control, which takes into account information about the application. It tries to give the user the necessary information in critical situations.

Although not explicitly mentioned in the literature, ergonomic rules and the user interface itself have to be considered. Any adaptation strategy or rule should regard ergonomic principles. The user interface as a whole has to be considered as the basis of any partial adaptation of the interface. Particularly, the adaptation strategy has to preserve the consistency of the user interface.

2.5 Timing of Adaptation

The timing strategies deal with the moment of changes in appearance and behavior of the user interface. Effects of timing are very striking for the user and for this reason enormously significant for the acceptance of the system by the user. [Rou88] mentions off-line adaptation prior to operation, on-line adaptation in anticipation of changing demands, and on-line adaptation in response to changes. [Coc87] merely distinguishes within-session and between-session adaptation. The following more detailed classification differentiates adaptation before the first utilization of the system, during use or between sessions.

Before first use of a system, several kinds of adaptation can take place. During the design of a user interface the needs of the expected typical user have to be considered. There are user interfaces which can be customized to the needs of special user groups before use. Obviously, the needs of the individual user cannot be considered in these cases. Some systems give the user the opportunity to tailor the system to personal preferences before use. This is not sufficient as needs appear during use. A customization with respect to the individual user happens, if the user is classified on the basis of a pre-test.

Perhaps the most interesting approach is termed *adaptation during use*. The adaptation can take place continuously, on predefined junctures, after (or before) defined functions, if a special situation appears, or on users' requests. Continuous adaptation can regard the current situation and the actual changes and has therefore the best chance to fit the user's needs. Furthermore, the results of the adaptation can be evaluated at once. Hence, a regressive adaptation can be obtained. On the other hand, an interface changing just that moment the user thought to understand it may cause major problems to casual users. Furthermore, adaptation during use can result in a so-called *hunting* [Bro90]: while the system tries to adapt to the user, the user tries to adapt to the system; they never reach a steady configuration.

Adaptation between two sessions allows to calculate very complicated adaptation strategies. The adaptation will always regard the users needs at the end of the last session. The conflict becomes obvious, if the user has not used the system for a long time.

3 Computer-Aided Adaptation

According to Chapter 2.2 a system incorporating Computer-Aided Adaptation can be described as follows. The user takes initiative for an adaptation. The system proposes different variants of sensible adaptations. The user can select a variant to be performed. If a proposed adaptation is selected by the user it has to be performed by the system.

The statements in this chapter additionally cover the systems with User-Controlled Adaptation. There is no significant difference concerning the basic principles of adaptation between this categories of systems, as they do only differ in the initiating agent.

The term *computer-aided* implies an adaptation fully controlled by the user. The system has to support the adaptation of the UI by the user. This includes the proposal of different variants if the user requests the adaptation of the interface. Furthermore, it will possibly include the ability to initiate the proposal of an adaptation if the system diagnoses the possibility of improving the communication by a change of the UI.

The users can select one of the proposed adaptations or reject them all. They can take initiative in the adaptation process and are free to decide for an adaptation not proposed by the system. They can propose different variants and are provided with an estimation of the expected effects on the whole interface.

An AUI offers three different types of meta dialog to the users providing the means for the adaptation. The users can change the UI by direct manipulation: They change the properties of UI objects, basic objects and combined objects in the same way.

The change by direct manipulation can be regarded as an implicit change of the user'spreferences concerning the UI. The second way for adaptation is the explicit change of user preferences by editing. Therefore it is necessary to provide a user model representing the

276

user preferences that is inspectable for the user. The system has to support the user by making available a browser with help or navigation facilities.

The third possibility for adapting the user interface is based on interview techniques. In order to build a user model as a basis for changes of the user interface the user is questioned by the system.

Besides an inspectable and editable user model, the major demands on Adaptive User Interfaces concern two aspects of compatibility. Firstly, it has to be considered that people are working together using the same system. It must be guaranteed that results obtained with a system adapted to the needs of a single user can be processed by a system adapted in a different way for a different user. This has to be warranted even if adaptation is applied to system functionality.

Secondly, it has to be considered that most users utilize different systems. All these systems should behave and react in a similar way. The adaptation of the user interface of a particular system should have consequences on the user interfaces of all the systems used by this user.

Concluding Remarks

This paper presents results of the SUITWARE project. The project is sited in the Department of System Ergonomics and Human-Computer Interaction at Corporate Research and Development of Siemens AG, Munich. The aspects of adaptivity presented in Chapter 2 are part of a taxonomy of AUI, available from the authors. The principles of Computer-Aided Adaptation described in Chapter 3 form the basis of the ongoing work on AUI in the SUITWARE project.

References

[Bro90]   D. Browne: *Conclusions*. In: [BTN90], pp. 195-212.

[BIM86]   D. Benyon, P. Innocent, D. Murray, J. Shergill: *Experiments in Adaptive Interfaces*. In: Proceedings IEE Colloquium on "Adaptive Man-Machine Interfaces" (Digest No 110), pp. 5/1-5/7, London, UK, 1986.

[BMJ90]   D. Benyon, D. Murray, F. Jennings: *An Adaptive System Developer's Toolkit*. In: D. Diaper, D. Gilmore, G. Cockton, B. Shackel (eds.): Human-Computer Interaction Interact'90, North-Holland, Amsterdam, 1990.

[BNR90]   D. Browne, M. Norman, D. Riches: *Why Build Adaptive Systems?*. In: [BTN90], pp. 15-57.

[BSN86]   D. P. Browne, B. D. Scharrat, M. A. Norman: *The Formal Specification of Adaptive User Interfaces Using Command Language Grammar*. SIGCHI Bulletin 17 (4), pp. 256-260, April 1986.

[BT88]   A. Brooks, C. Thorburn: *User-driven Adaptive Behaviour, A Comparative Evaluation And An Inductive Analysis*. In: D. Jones, R. Winder (Eds.): People and computers IV. Proceedings of the Fourth Conference of the British Computer Society (Univ. of Manchester, UK, Sept. 5-6, 1988), pp. 237-255. Cambridge University Press, New York, 1988.

[BTN90]   D. Browne, P. Totterdell, M. Norman (Eds.): *Adaptive User Interfaces*. Academic Press, London, 1990.

[Coc87]   G. Cockton: *Some Critical Remarks on Abstractions for Adaptable Dialogue Managers*. In: D. Diaper, R. Winder (Eds.): People and Computers III. Proceedings of the Third Conference of the British Computer Society, Human Computer Interaction Specialist Group, University of Exeter, Sept. 7-11, 1987, pp. 325-343.

[Cot90]   J. A. Cote Muñoz: AIDA - *Ein an den Benutzer angepaßtes Graphisch-Interaktives System*. Doctoral Thesis (german). Darmstädter Dissertation D17, TH Darmstadt, 1990.

[Cro84]   W. B. Croft: *The role of context and adaptation in user interfaces*. Int. J. Man-Machine Studies 21, pp. 283-292, 1984.

[Dan88]   W. Dang: *Intelligence in a User Interface Management System*. ERGO-IA '88. European Colloquium - Ergonomics and Artificial Intelligence. Proceedings. CNRS, Univ. de Paris-Sud, Orsay, France, 1988.

[EW89]   J. Elkerton, R. C. Williges: *Dialogue Design for Intelligent Interfaces*. In: P. A. Hancock, M. H. Chignell (Eds.): Intelligent Interfaces: Theory, Research and Design. North-Holland, Amsterdam, 1989, pp. 213-264.

[FMS87]   C. J. H. Fowler, L. A. Macaulay, S. Siripoksup: *An Evaluation of the Effectiveness of the Adaptive Interface Module (AIM) in Matching Dialogues to Users*. In: D. Diaper, R. Winder (Eds.): People and Computers III. Proceedings of the Third Conference of the British Computer Society, Human Computer Interaction Specialist Group, University of Exeter, Sept. 7-11, 1987, pp. 345-359.

[FWK91]   J. Foley, Won Chul Kim, S. Kovacevic, K. Murray: *UIDE - An Intelligent User Interface Design Environment*. In: J. W. Sullivan, S. W. Tyler (Eds.): Intelligent User Interfaces. ACM Press, New York, 1991, pp. 339-384.

[Gri86]   R. Grimm: *ADBS: A Tool for Designing and Implementing the Man-Process Interface for Different Users*. In: Analysis, Design and Evaluation of Man-Machine Systems, pp. 287-291. Proceedings of the 2nd IFAC/IFIP/IFORS/IEA Conference, Oxford, England, 1986.

[HP91]   H. U. Hoppe, R. Plötzner: *Inductive Knowledge Acquisition for a UNIX Coach*. In: M. J. Tauber, D. Ackermann (Eds.): Mental Models and Human-Computer Interaction 2, pp. 313-335. Elsevier, Amsterdam, 1991.

[Mas86]   M. V. Mason: *Adaptive command prompting in an on-line documentation system*. Int. J. Man-Machine Studies 25, pp. 33-51, 1986.

278

[Mot90]   A. Motro: *Flex: A Tolerant and Cooperative User Interface to Databases.* IEEE Transactions on Knowledge and Data Engineering 2 (2), pp. 231-246, June 1990

[MRW86]   N. M. Morris, W. B. Rouse, S. L. Ward: *Experimental Evaluation of Adaptive Task Allocation in an Aerial Search Environment.* In: Analysis, Design and Evaluation of Man-Machine Systems, pp. 67-72. Proceedings of the 2nd IFAC/IFIP/IFORS/IEA Conference, Oxford, England, 1986.

[Ris84]   E. L. Rissland: *Ingredients of intelligent user interfaces.* Int. J. Man-Machine Studies 21, pp. 377-388, 1984.

[Rou88]   W. B. Rouse: *Adaptive Aiding for Human/Computer Control.* Human Factors 30 (4), pp. 431-443, 1988

[Sch89]   T. Schwab: *Methoden zur Dialog- und Benutzermodellierung in adaptiven Computersystemen.* Doctoral Thesis (german). Institut für Informatik, Universität Stuttgart, 1989.

[She90]   E. H. Sherman: *A User-Adaptable Interface to Predict Users' Needs.* Knowledge Systems Laboratory, Report KSL-90-56, Medical Computer Science, Stanford University, August 1990.

[SB82]   D. Sleeman, J.S. Brown: *Introduction: Intelligent tutoring systems.* In: D. Sleeman, J.S. Brown (Eds.): Intelligent Tutoring Systems. pp. 1-11, Academic Press, London, 1982.

[TR90]   P. Totterdell, P. Rautenbach: *Adaptation as a Problem of Design.* In: [BTN90], pp. 59-84, 1990.

[TT89]   S. W. Tyler, S. Treu: *An interface architecture to provide adaptive task-specific context for the user.* Int. J. Man-Machine Studies 30, pp. 303-327, 1989.

[VB88]   G. C. van der Veer, J. J. Beishuizen: *Computers and Education: Adaptation to Individual Differences.* In G. C. van der Veer, T. R. G. Green, J.-M. Hoc, D. M. Murray (Eds.): Working with Computers: Theory versus Outcome. Academic Press, London, 1988, pp. 251-278.

[WAC84]   R. Wilensky, Y. Arens, D. N. Chin: *Talking to UNIX in English: An Overview of UC.* Comm. of the ACM 27, pp. 574-593, 1984.

Workstations and HDTV

Walter Woborschil
Siemens AG

Abtract

Multimedia is currently one of the main goals of workstation design. At the same time High Definition Television (HDTV) is in worldwide development. This paper will give a survey over techniques, standardization, technology, and applications of HDTV:

- HDTV is leading the way towards an all digital television. Fully digital HDTV systems show lot of synergies with multimedia PCs/workstations.
- HDTV drives the development of new technologies in microelectronics and display technology.
- HDTV sets and PCs/workstations will increasingly share a common market, possibly leading to new kind of computer systems.

As digital technology is the main part of HDTV, application possibilities go far beyond conventional television. Especially the interconnection with, and the impact on workstations are addressed.

1 Introduction

High Definition Television (HDTV) stands for a high quality television technique. Image quality is improved by about twice the number of display lines and about twice the number of pixels per line compared to standard television. In addition an HDTV objective is larger image size (larger than 40 inches) with a horizontal to vertical ratio of 16 to 9 to improve the human image perception.

First HDTV developments started about 20 years ago. The early goal was to develop a next generation television system mainly based on existing television technology. Systems were characterized by analog transmission and some digital image processing. Rapid advances in digital technologies, especially in digital communication and digital image compression allow future all digital HDTV systems. Currently some fully digital HDTV approaches are already under evaluation in the USA. Fully digital HDTV systems show similarities to data processing systems, in particular workstations. Exploiting these synergies may result in future information processing systems having new functionality like multimedia workstations with HDTV image resolution or HDTV sets with added PC functionality.

2 HDTV Applications

Traditional applications of HDTV are broadcast applications within the consumer market as well as in television studios. New application possibilities emerge in the nonbroadcast area, see fig. 1.

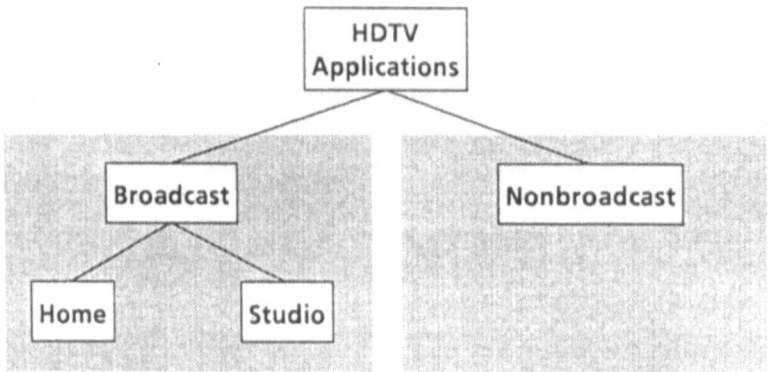

Figure 1: HDTV application areas

Concerning broadcast applications HDTV sets targeted for the consumer market offer larger image sizes, higher image resolution, and improved color and audio quality. For television studios HDTV offers the possibility to define a world-wide standard for high resolution video leading to simplified international television program exchange. However current developments are towards three different standards in the USA, in Japan and in Europe.

New application potential is in the nonbroadcast area, especially for applications where high resolution imaging or image communication is required like desktop publishing, computer aided design, medical image processing and archiving, visualization of scientific data, and teleconferencing.

3 HDTV Standards

The development of the japanese "High Vision" HDTV system has started about 20 years ago. At that time image coding was only feasable by subsampling algorithms. MUSE, the "High Vision" coding algorithm yields a bandwidth reduction of about a factor of 4. The effective horizontal resolution is about 1440 pixel, the vertical resolution is 1035 lines with interlaced scanning at a field rate of 60 Hz. MUSE uses analog signal transmission via direct broadcasting satellites. Since 1989 regular test transmissions are carried out. For the "High Vision" system television sets as well as complete studio equipment including cameras and recorders are commercially available.

In the middle of the eighties the european HDTV initiative was launched with the EUREKA 95 project. The HD-MAC coding algorithm uses adaptive subsampling and is upwards compatible to D2-MAC, the already existing european standard for broadcast satellite transmission. The effective resolution at the receiver is about 1440 pixels by 1152 lines. The EU 95 studio format defines a resolution of 1920 by 1152 pixel. Interlaced scanning at a field rate of 50 Hz is used. Market introduction is planned to be in 1995.

In the USA HDTV development started just within the last few years. Currently six different HDTV proposals have been submitted to the FCC and are evaluated within a field trial. A decision about which system to choose is planned to be made by the end of 1993. Two of the proposals are still based on a "conventional" subsampling and analog transmission approach. However four of the proposals already promote fully digital systems including digital terrestrial transmission within available television channels (simulcast). This requires digital image compression by a factor of about 30. The compression techniques applied are quite similar to techniques used in video telephony (H.261) and multimedia (MPEG). The all digital proposals are DSC-HDTV (Zenith/AT&T), ATVA Progressive, ATVA Interlace (General Instruments, MIT), and ADTV (ATRC). These all digital proposals are regarded as the more advanced systems. Features like "progressive scan" and "square pixels" , which are essential to ease the processing of video pictures in a computer environment, are addressed.

Figure 2 gives an overview of the different HDTV approaches.

Japan	Europe	USA
High Vision / MUSE	EU 95 / HDMAC	6 proposals
• Analog system digital image processing	• Analog system digital image processing	• 4 proposals fully digital, including transmission
• 1920x1035, 60Hz	• 1920x1152, 50Hz	• c. 1300x1000, 60Hz
• Ready for market	• Introduction: 1992	• Decision on final system: 1993
• Regular pilot transmissions	• Put on market: 1994	• Simulcast
• Complete studio equipment available	• Transmission via satellite	
• Transmission via satellite	• First work on fully digital systems	
• First work on fully digital systems		

Figure 2: Emerging HDTV standards

Currently it seems that no world-wide standard can be achieved. Fully digital systems will become more and more significant in the future.

4 Workstation Trends

Besides the general trends towards increasing performance (Joy's law predicts doubling every year; Bill Joy, SUN Microsystems), cost reduction and physical smaller systems, future workstations will have additional advanced features. This will make the use of workstations in new application areas feasible.

About a decade ago where workstations were introduced, the main features were a processor having 1 Mips performance, a graphical display with about 1 million pixels, and the interconnection by a local area network. An important key element for the success of workstations were graphical user interfaces, e.g. the X-Window system. Within the last years 2D graphics has been expanded to 3D graphics. High performance workstations now allow the real-time animation of complex 3D scenes. Applications which have been opened up are e.g. CAD, visualization of scientific data, and virtual reality. The next step in workstation development are multimedia workstations, this means workstations which additionally can handle audio and live video information. Since the introduction of workstations this will have the major impact on new application areas. A smooth integration of audio and video information into workstations requires fully digital representation. The main challenge for the workstation designer is the integration of live video. To reduce the amount of real-time data image compression techniques are necessary. However extremely high processing performance in the range of Gops is required. Promising new applications of multimedia workstations are teleconferencing and computer supported cooperative work. Broadband networking will enable communication over wide area networks with transmission speeds of about 100 Mbit/s. In addition mobile communication will allow a more flexible use of portable systems.

Workstations will evolve from dedicated systems used by engineers to tools used by professionals. This requires also more "intelligent" systems, which assist their users in daily work.

5 Synergies between HDTV and Workstations

The improvements of HDTV are mainly achieved by use of digital signal processing. With the computer industry heading towards multimedia and live video on desktop systems, the digital processing of video sequences that comes with HDTV will support the integration of high quality video into future workstations. Some workstation manufactures are currently working on prototypes of HDTV workstations.

The features of HDTV attracting the computer industry are not mainly higher resolution of images and displays, but the fact, that digital processing is a main characteristic of HDTV. The large bandwidth that is required by HDTV signals make data compression inevitable for economic storage and transmission. Compression of video data is therefore one of the main links between HDTV and multimedia workstations. On the one hand a common compression standard leads to compatible equipment and video program material between HDTV and multimedia systems, on the other hand it facilitates the economic development of VLSI compression hardware.

Digital processing offers new and more comfortable ways for image processing to television and film studios, too. Computers will be an important part in these studios. As technology develops todays high-end equipment becomes tomorrows standard equipment. Instead of HDTV sets with mere reproduction abilities more sophisticated "video stations", enriched with additional interfaces like keyboard and camera, may provide the same level of functionality as a multimedia workstation (see fig. 3).

284

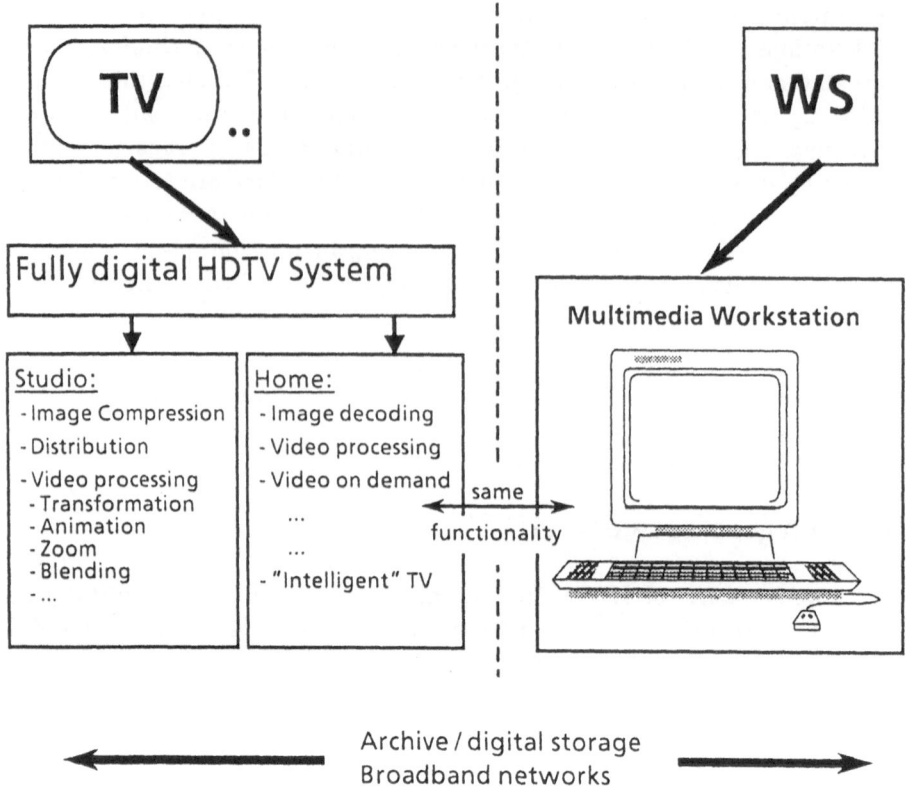

Figure 3: Synergies between HDTV and Multimedia

Therefore a convergence of the consumer electronics industry offering HDTV and the computing industry offering multimedia systems is expected.

6 Conclusion

By HDTV a new television technique having significantly improved video and audio quality is developed. In Europe and Japan HDTV systems are only partly based on digital signal processing. Fully digital systems including digital transmission are regarded as a further step of development. In the USA it is attempted to achieve an all digital system within one step of development.

Fully digital HDTV systems show similarities to multimedia workstations or PCs. That concerns the processing of video information within television studios and television sets, necessary compression techniques for transmission and storage of video information, and the required media for transmission and storage. The high demands referring to image quality as well as the big market volume expected make HDTV a technology driver for microelectronics, display, storage, and transmission techniques where the data processing industry can take part in. Examples are the intensified development of big-sized flat screens and high performance special processors for image processing and compression.

There doesn't appear a world-wide uniform HDTV standard to be established in the near future. While current HDTV standards in Europe and Japan, which are for the most part based on analog techniques, will hardly affect data processing, an all digital standard as proposed in the USA may also be the basis for a standard for live video handling within computer systems. Another advantage is the common usage of video archives and system components. A fully digital HDTV is also suitable for transmission within BISDN networks which offer new areas of applications like video on demand services.

Further applications of HDTV arise at the qualitative improvement of current multimedia applications and of video communication. By intensive usage of digital techniques and the integration of PC or workstation functionality, these capabilities can also be offered by future HDTV receivers. This will result in an overlapping of the market segments for forthcoming HDTV receiver and for PC/workstations.

References

[1] "Rivista Telettra Review 45", Special Issue for High Definition TV, Telettra, (1990), ISSN 0392-8268

[2] "Specifications of the HDMAC/packet System", European Telecommunications Standards Institute (1991)

[3]   Y.Ninomiya et al: "An HDTV Broadcast System Utilizing a Bandwidth Compression Technique - MUSE", IEEE Transactions on Broadcasting No 4 (1987)

[4]   R.M. Rast: "The Alliance Interlace HDTV System", FCC/ACA proposal, General Instruments (1991)

[5]   "ATVA-Progressive system", FCC/ACA proposal, MIT, Cambridge MA (1991)

[6]   "Technical Description: Digital Spectrum Compatible HDTV system", FCC/ACA proposal, Zenith Electronics, AT&T (1991)

[7]   "Advanced Digital Television System Description", FCC/ACA proposal, David Sarnoff Research Center, NBC, Philips N.A., Thomson (1991)

[8]   R. E. Keeler, "Interoperability Considerations for Digital HDTV", IEEE Transactions on Broadcasting, Vol. 37, No. 4 (1991)

[9]   "TV of Tomorrow", MIT Media Laborotory, Cambridge MA (1990)

[10] Didier Le Gall, "MPEG: A Video Compression Standard for Multimedia Applications", Communications of the ACM, Vol. 34, No. 4 (1991)

[11] C. Müller-Schloer, E.Schmitter: "RISC-Workstation-Architekturen", Springer Verlag Berlin (1991)

[12] H.Johnen, M.Östreicher, W. Woborschil: "HDTV und Datenverarbeitung", Siemens Internal Report, ZFE ST SN 1.141, (1992)